Springer Texts in
Electrical Engineering

Springer Texts in Electrical Engineering

Multivariable Feedback Systems
F.M. Callier/C.A. Desoer

Linear Programming
M. Sakarovitch

Introduction to Random Processes
E. Wong

Stochastic Processes in Engineering Systems
E. Wong/B. Hajek

Introduction to Probability
J.B. Thomas

Elements of Detection and Signal Design
C.L. Weber

An Introduction to Communication Theory and Systems
J.B. Thomas

Signal Detection in Non-Gaussian Noise
S.A. Kassam

Saleem A. Kassam

Signal Detection in Non-Gaussian Noise

Consulting Editor: John B. Thomas

With 38 Illustrations

Springer-Verlag
New York Berlin Heidelberg
London Paris Tokyo

A Dowden &
Culver Book

Saleem A. Kassam
Department of Electrical Engineering
University of Pennsylvania
Philadelphia, PA 19104
USA

Library of Congress Cataloging-in-Publication Data
Kassam, Saleem A.
  Signal detection in non-Gaussian noise.
  (Springer texts in electrical engineering)
  Bibliography: p.
  Includes index.
  1. Signal theory (Telecommunication) 2. Signal
detection. I. Title. II. Series.
  TK5102.5.K357 1987   621.38′0436   87-32197

© 1988 by Dowden & Culver, Inc.
All rights reserved. This work may not be translated or copied in whole or in part without the written permission of the copyright holder, except for brief excerpts in connection with reviews or scholarly analysis. Use in connection with any form of information storage and retrieval, electronic adaptation, computer software, or by similar or dissimilar methodology now known or hereafter developed is forbidden.
The use of general descriptive names, trade names, trademarks, etc. in this publication, even if the former are not especially identified, is not to be taken as a sign that such names, as understood by the Trade Marks and Merchandise Marks Act, may accordingly be used freely by anyone.

Camera-ready copy provided by the author.
Printed and bound by R.R. Donnelley & Sons, Harrisonburg, Virginia.
Printed in the United States of America.

9 8 7 6 5 4 3 2 1

ISBN 0-387-96680-3 Springer-Verlag New York Berlin Heidelberg
ISBN 3-540-96680-3 Springer-Verlag Berlin Heidelberg New York

PREFACE

This book contains a unified treatment of a class of problems of signal detection theory. This is the detection of signals in additive noise which is not required to have Gaussian probability density functions in its statistical description. For the most part the material developed here can be classified as belonging to the general body of results of *parametric* theory. Thus the probability density functions of the observations are assumed to be known, at least to within a finite number of unknown parameters in a known functional form. Of course the focus is on noise which is not Gaussian; results for Gaussian noise in the problems treated here become special cases. The contents also form a bridge between the classical results of signal detection in Gaussian noise and those of *nonparametric* and *robust* signal detection, which are not considered in this book.

Three canonical problems of signal detection in additive noise are covered here. These allow between them formulation of a range of specific detection problems arising in applications such as radar and sonar, binary signaling, and pattern recognition and classification. The simplest to state and perhaps the most widely studied of all is the problem of detecting a *completely known deterministic signal* in noise. Also considered here is the detection of a *random non-deterministic signal* in noise. Both of these situations may arise for observation processes of the low-pass type and also for processes of the band-pass type. Spanning the gap between the known and the random signal detection problems is that of detection of a deterministic signal with random parameters in noise. The important special case of this treated here is the detection of *phase-incoherent narrowband signals* in narrowband noise.

There are some specific assumptions that we proceed under throughout this book. One of these is that ultimately all the data which our detectors operate on are *discrete sequences* of observation components, as opposed to being continuous-time waveforms. This is a reasonable assumption in modern implementations of signal detection schemes. To be able to treat non-Gaussian noise with any degree of success and obtain explicit, canonical, and useful results, a more stringent assumption is needed. This is the *independence* of the discrete-time additive noise components in the observation processes. There do exist many situations under which this assumption is at least a good approximation.

With the same objective of obtaining explicit canonical results of practical appeal, this book concentrates on *locally optimum* and *asymptotically optimum* detection schemes. These criteria are appropriate in detection of weak signals (the low signal-to-noise-ratio case), for which the use of optimum detectors is particularly meaningful and necessary to extract the most in detection performance.

Most of the development given here has not been given detailed exposition in any other book covering signal detection theory and applications, and many of the results have appeared relatively recently in technical journals. In presenting this material it is assumed only that the reader has had some exposure to the elements of statistical inference and of signal detection in Gaussian noise. Some of the basic statistics background needed to appreciate the rest of the development is reviewed in Chapter 1. This book should be suitable for use in a first graduate course on signal detection, to supplement the classical material on signal detection in Gaussian noise. Chapters 2-4 may be used to provide a fairly complete introduction to the known signal detection problem. Chapters 5 and 6 are on the detection of narrowband known and phase-incoherent signals, respectively, and Chapter 7 is on random signal detection. A more advanced course on signal detection may also be based on this book, with supplementary material on nonparametric and robust detection if desired. This book should also be useful as a reference to those active in research, as well as to those interested in the application of signal detection theory to problems arising in practice.

The completion of this book has been made possible through the understanding and help of many individuals. My family has been most patient and supportive. My graduate students have been very stimulating and helpful. Prashant Gandhi has been invaluable in getting many of the figures ready. For the excellent typing of the drafts and the final composition, I am grateful to Drucilla Spanner and to Diane Griffiths. Finally, I would like to acknowledge the research support I have received from the Air Force Office of Scientific Research and the Office of Naval Research, which eventually got me interested in writing this book.

Saleem A. Kassam
University of Pennsylvania
Philadelphia, PA 19104

TABLE OF CONTENTS

	Page
Chapter 1 Elements of Statistical Hypothesis Testing	1
1.1 Introduction	1
1.2 Basic Concepts of Hypothesis Testing	2
1.3 Most Powerful Tests and the Neyman-Pearson Lemma	3
1.4 Local Optimality and the Generalized Neyman-Pearson Lemma	5
1.5 Bayes Tests	8
1.6 A Characterization of the Relative Performance of Tests	10
1.6.1 Relative Efficiency	11
1.6.2 Asymptotic Relative Efficiency and Efficacy	12
Problems	21
Chapter 2 Detection of Known Signals in Additive Noise	24
2.1 Introduction	24
2.2 The Observation Model	24
2.2.1 Detection of a Pulse Train	25
2.2.2 Statistical Assumptions	29
2.3 Locally Optimum Detection and Asymptotic Optimality	31
2.3.1 Locally Optimum Detectors	34
2.3.2 Generalized Correlator Detectors	37
2.3.3 Asymptotic Optimality	39
2.3.4 Asymptotic Optimality of LO Detectors	42
2.4 Detector Performance Comparisons	46
2.4.1 Asymptotic Relative Efficiency and Efficacy	47
2.4.2 Finite-Sample-Size Performance	54
2.5 Locally Optimum Bayes Detection	64
2.6 Locally Optimum Multivariate Detectors	67
Problems	70

Chapter 3 Some Univariate Noise Probability Density Function Models — 72

3.1 Introduction — 72
3.2 Generalized Gaussian and Generalized Cauchy Noise — 73

    3.2.1 Generalized Gaussian Noise — 74
    3.2.2 Generalized Cauchy Noise — 78

3.3 Mixture Noise and Middleton Class A Noise — 84
3.4 Adaptive Detection — 91

Problems — 94

Chapter 4 Optimum Data Quantization in Known-Signal Detection — 97

4.1 Introduction — 97
4.2 Asymptotically Optimum Quantization — 99
4.3 Asymptotically Optimum Generalized Quantization — 111
4.4 Maximum-Distance Quantization — 115
4.5 Approximations of Locally Optimum Test Statistics — 118

Problems — 124

Chapter 5 Detection of Known Narrowband Signals in Narrowband Noise — 127

5.1 Introduction — 127
5.2 The Observation Model — 128
5.3 Locally Optimum Detection — 131
5.4 Asymptotic Performance Analysis — 139
5.5 Asymptotically Optimum Envelope Quantization — 144
5.6 Locally Optimum Bayes Detection — 146

Problems — 148

Chapter 6 Detection of Narrowband Signals with Random Phase Angles — 151

6.1 Introduction — 151
6.2 Detection of an Incoherent Signal — 151

	6.2.1 The Observation Model	152
	6.2.2 The Locally Optimum Detector	153
6.3	Detection of a Noncoherent Pulse Train	162
	6.3.1 The Observation Model	163
	6.3.2 The Locally Optimum Detector	166
6.4	Asymptotically Optimum Quantization	172
	6.4.1 M-Interval Envelope Quantization	172
	6.4.2 Optimum Envelope Quantization for Gaussian Noise	175
	6.4.3 M-Region Generalized Quantization	181
Problems		183

Chapter 7 Detection of Random Signals in Additive Noise — 185

7.1	Introduction	185
7.2	The Observation Model	186
7.3	Locally Optimum Array Detection	188
	7.3.1 General Form of the LO Test Statistic	188
	7.3.2 LO Statistic for White Signal Sequence	192
	7.3.3 LO Statistic for Correlated Signal Sequence	196
7.4	Asymptotic Performance Characteristics	198
7.5	Asymptotically Optimum Quantization	204
7.6	Detection of Narrowband Random Signals	208
Problems		211
References		215
Index		227

Chapter 1

ELEMENTS OF STATISTICAL HYPOTHESIS TESTING

1.1 Introduction

The signal processing problem which is the object of our study in this book is that of detecting the presence of a signal in noisy observations. Signal detection is a function that has to be implemented in a variety of applications, the more obvious ones being in radar, sonar, and communications. By viewing signal detection problems as problems of binary hypothesis testing in statistical inference, we get a convenient mathematical framework within which we can treat in a unified way the analysis and synthesis of signal detectors for different specific situations. The theory and results in mathematical statistics pertaining to binary hypothesis-testing problems are therefore of central importance to us in this book. In this first chapter we review some of these basic statistics concepts. In addition, we will find in this chapter some further results of statistical hypothesis testing with which the reader may not be as familiar, but which will be of use to us in later chapters.

We begin in Section 1.2 with a brief account of the basic concepts and definitions of hypothesis-testing theory, which leads to a discussion of most powerful tests and the Neyman-Pearson lemma in Section 1.3. In Section 1.4 this important result is generalized to yield the structures of locally optimum tests, which we will make use of throughout the rest of this book. Section 1.5 reviews briefly the Bayesian approach to construction of tests for hypotheses. We shall not be using the Bayesian framework very much except in Chapters 2 and 5, where we shall develop locally optimum Bayes' detectors for detection of known signals in additive noise.

In the last section of this chapter we will introduce a measure which we will make use of quite extensively in comparing the performances of different detectors for various signal detection problems in the following chapters. While we will give a more general discussion of the *asymptotic relative efficiency* and the *efficacy* in Section 1.6, these measures will be introduced and discussed in detail for the specific problem of detection of a known signal in additive noise in Section 2.4 of Chapter 2. Readers may find it beneficial to postpone study of Section 1.6 until after Chapter 2 has been read; they may then better appreciate the applicability of the ideas and results of this section.

1.2 Basic Concepts of Hypothesis Testing

Let $\mathbf{X} = (X_1, X_2, ..., X_n)$ be a random vector of observations with joint probability density function (pdf) $f_{\mathbf{X}}(\mathbf{x} \mid \theta)$, where θ is a parameter of the density function. Any specific realization $\mathbf{x} = (x_1, x_2, ..., x_n)$ of $\mathbf{X}$ will be a point in $\mathbb{R}^n$, where $\mathbb{R}$ is the set of all real numbers. In binary hypothesis-testing problems we have to decide between one of two hypotheses, which we will label as H and K, about the pdf $f_{\mathbf{X}}(\mathbf{x} \mid \theta)$, given an observation vector in $\mathbb{R}^n$. Let Θ be the set of all possible values of θ; we usually identify H with one subset Θ_H of θ values and K with a disjoint subset Θ_K, so that $\Theta = \Theta_H \cup \Theta_K$. This may be expressed formally as

$$H: \mathbf{X} \text{ has pdf } f_{\mathbf{X}}(\mathbf{x} \mid \theta) \text{ with } \theta \in \Theta_H \qquad (1\text{-}1)$$

$$K: \mathbf{X} \text{ has pdf } f_{\mathbf{X}}(\mathbf{x} \mid \theta) \text{ with } \theta \in \Theta_K \qquad (1\text{-}2)$$

If Θ_H and Θ_K are made up of single elements, say θ_H and θ_K, respectively, we say that the hypotheses are *simple*; otherwise, the hypotheses are *composite*. If Θ can be viewed as a subset of $\mathbb{R}^p$ for a finite integer p, the pdf $f_{\mathbf{X}}(\mathbf{x} \mid \theta)$ is completely specified by the finite number p of real components of θ, and we say that our hypotheses are *parametric*.

A *test* for the hypothesis H against K may be specified as a partition of the sample space $S = \mathbb{R}^n$ of observations into disjoint subsets S_H and S_K, so that $\mathbf{x}$ falling in S_H leads to acceptance of H, with K accepted otherwise. This may also be expressed by a *test function* $\delta(\mathbf{x})$ which is defined to have value $\delta(\mathbf{x}) = 1$ for $\mathbf{x} \in S_K$ and value $\delta(\mathbf{x}) = 0$ for $\mathbf{x} \in S_H$. The value of the test function is defined to be the probability with which the hypothesis K, the *alternative hypothesis*, is accepted. The hypothesis H is called the *null hypothesis*.

More generally, the test function can be allowed to take on probability values in the closed interval [0,1]. A test based on a test function taking on values inside [0,1] is called a *randomized test*.

The *power function* $p(\theta \mid \delta)$ of a test based on a test function δ is defined for $\theta \in \Theta_H \cup \Theta_K$ as

$$\begin{aligned} p(\theta \mid \delta) &= E\{\delta(\mathbf{X}) \mid \theta\} \\ &= \int_{\mathbb{R}^n} \delta(\mathbf{x}) f_{\mathbf{X}}(\mathbf{x} \mid \theta) \, d\mathbf{x} \end{aligned} \qquad (1\text{-}3)$$

Thus it is the probability with which the test will accept the alternative hypothesis K for any particular parameter value θ. When

θ is in Θ_H the value of $p(\theta\mid\delta)$ gives the probability of an error, that of accepting K when H is correct. This is called a *type I error*, and depends on the particular value of θ in Θ_H. The *size* of a test is the quantity

$$\alpha = \sup_{\theta\in\Theta_H} p(\theta\mid\delta) \tag{1-4}$$

which may be considered as being the best upper bound on the type I error probability of the test.

In signal detection the null hypothesis is often a noise-only hypothesis, and the alternative hypothesis expresses the presence of a signal in the observations. For a detector D implementing a test function $\delta(\mathbf{x})$ the power function evaluated for any θ in Θ gives a *probability of detection* of the signal. Thus in later chapters we will use the notation $p_d(\theta\mid D)$ for the power function of a detector D, and in discussing the probability of detection at a particular value of the parameter θ in Θ_K (or for a simple alternative hypothesis K) we will use for it the notation p_d. The size of a detector is often called its *false-alarm probability*. This usage is encountered specifically when the noise-only null hypothesis is simple, and the notation for this probability is p_f.

1.3 Most Powerful Tests and the Neyman-Pearson Lemma

Given a problem of binary hypothesis testing such as defined by (1-1) and (1-2), the question arises as to how one may define and then construct an optimum test. Ideally, one would like to have a test for which the power function $p(\theta\mid\delta)$ has values close to zero for θ in Θ_H, and has values close to unity for θ in Θ_K. These are, however, conflicting requirements. We can instead impose the condition that the size α of any acceptable test be no larger than some reasonable level α_0, and subject to this condition look for a test for which $p(\theta\mid\delta)$, evaluated at a particular value θ_K of θ in Θ_K, has its largest possible value. Such a test is *most powerful* at level α_0 in testing H against the simple alternative $\theta = \theta_K$ in Θ_K; its test function $\delta^*(\mathbf{x})$ satisfies

$$\sup_{\theta\in\Theta_H} p(\theta\mid\delta^*) \leq \alpha_0 \tag{1-5}$$

and

$$p(\theta_K\mid\delta^*) \geq p(\theta_K\mid\delta) \tag{1-6}$$

for all other test functions $\delta(\mathbf{x})$ of size less than or equal to α_0. In most cases of interest a most powerful level α_0 test satisfies (1-5) with equality, so that its size is $\alpha = \alpha_0$.

For a simple null hypothesis H when $\theta = \theta_H$ is the only parameter value in Θ_H, the condition (1-5) becomes $p(\theta_H \mid \delta^*) \leq \alpha_0$ or $p_f \leq \alpha_0$, subject to which p_d at $\theta = \theta_K$ is maximized. For this problem of testing a simple H against a simple K, a fundamental result of Neyman and Pearson (called the Neyman-Pearson lemma) gives the structure of the most powerful test. We state the result here as a theorem:

Theorem 1: Let $\delta(\mathbf{x})$ be a test function of the form

$$\delta(\mathbf{x}) = \begin{cases} 1 & , \quad f_{\mathbf{x}}(\mathbf{x} \mid \theta_K) > t f_{\mathbf{x}}(\mathbf{x} \mid \theta_H) \\ r(\mathbf{x}) & , \quad f_{\mathbf{x}}(\mathbf{x} \mid \theta_K) = t f_{\mathbf{x}}(\mathbf{x} \mid \theta_H) \\ 0 & , \quad f_{\mathbf{x}}(\mathbf{x} \mid \theta_K) < t f_{\mathbf{x}}(\mathbf{x} \mid \theta_H) \end{cases} \qquad (1\text{-}7)$$

for some constant $t \geq 0$ and some function $r(\mathbf{x})$ taking on values in [0,1]. Then the resulting test is most powerful at level equal to its size for $H: \theta = \theta_H$ versus $K: \theta = \theta_K$.

In addition to the above sufficient condition for a most powerful test it can be shown that conversely, if a test is known to be most powerful at level equal to its size, then its test function must be of the form (1-7) except perhaps on a set of $\mathbf{x}$ values of probability measure zero. Additionally, we may always require $r(\mathbf{x})$ in (1-7) to be a constant r in [0,1]. Finally, we note that we are always guaranteed the existence of such a test for H versus K, of given size α [Lehmann, 1959, Ch. 3].

From the above result we see that generally the structure of a most powerful test may be described as one comparing a *likelihood ratio* to a constant *threshold*,

$$\frac{f_{\mathbf{x}}(\mathbf{x} \mid \theta_K)}{f_{\mathbf{x}}(\mathbf{x} \mid \theta_H)} > t \qquad (1\text{-}8)$$

in deciding if the alternative K is to be accepted. If the likelihood ratio on the left-hand side of (1-8) equals the threshold value t, the alternative K may be accepted with some probability r (the randomization probability). The constants t and r may be evaluated to obtain a desired size α using knowledge of the distribution function of the likelihood ratio under H.

When the alternative hypothesis K is composite we may look for a test which is *uniformly most powerful* (UMP) in testing H against K, that is, one which is most powerful for H against each $\theta = \theta_K$ in Θ_K. While UMP tests can be found in some cases, notably in many situations involving Gaussian noise in signal detection, such tests do not exist for many other problems of interest. One option in such situations is to place further restrictions on the class of acceptable or admissible tests in defining a most powerful test; for example, a requirement of *unbiasedness* or of *invariance* may be imposed [Lehmann, 1959, Ch. 4-6]. As an alternative, other performance criteria based on the power function may be employed. We will consider one such criterion, leading to *locally optimum* or *locally most powerful* tests for composite alternatives, in the next section. One approach to obtaining reasonable tests for composite hypotheses is to use maximum-likelihood estimates $\hat{\theta}_H$ and $\hat{\theta}_K$ of the parameter θ, obtained under the constraints $\theta \in \Theta_H$ and $\theta \in \Theta_K$, respectively, in place of θ_H and θ_K in (1-8). The resulting test is called a *generalized likelihood ratio* test or simply a *likelihood ratio* test (see, for example, [Bickel and Doksum, 1977, Ch. 6]).

1.4 Local Optimality and the Generalized Neyman-Pearson Lemma

Let us now consider the approach to construction of tests for composite alternative hypotheses which we will use almost exclusively in the rest of our development on signal detection in non-Gaussian noise. In this approach attention is concentrated on alternatives $\theta = \theta_K$ in Θ_K which are close, in the sense of a metric or distance, to the null-hypothesis parameter value $\theta = \theta_H$. Specifically, let θ be a real-valued parameter with value $\theta = \theta_0$ defining the simple null hypothesis and let $\theta > \theta_0$ define the composite alternative hypothesis. Consider the class of all tests based on test functions $\delta(\mathbf{x})$ of a particular desired size α for $\theta = \theta_0$ against $\theta > \theta_0$, and assume that the power functions $p\,(\theta \mid \delta)$ of these tests are continuous and also continuously differentiable at $\theta = \theta_0$. Then if we are interested primarily in performance for alternatives which are close to the null hypothesis, we can use as a measure of performance the slope of the power function at $\theta = \theta_0$, that is,

$$p'\,(\theta_0 \mid \delta) = p'\,(\theta \mid \delta)\,|_{\theta = \theta_0}$$
$$= \frac{d}{d\theta}\,p\,(\theta \mid \delta)\bigg|_{\theta = \theta_0} \qquad (1\text{-}9)$$

From among our class of tests of size α, the test based on $\delta^*(\mathbf{x})$ which uniquely maximizes $p'\,(\theta_0 \mid \delta)$ has a power function satisfying

$$p\,(\theta \mid \delta^*) \geq p\,(\theta \mid \delta), \quad \theta_0 < \theta < \theta_{\max} \qquad (1\text{-}10)$$

for some $\theta_{max} > \theta_0$. Such a test is called a *locally* most powerful or *locally optimum* (LO) test for $\theta = \theta_0$ against $\theta > \theta_0$. It is clearly of interest in situations such as the weak-signal case in signal detection, when the alternative-hypothesis parameter values of primary concern are those which define pdf's $f_X(x \mid \theta)$ close to the null-hypothesis noise-only pdf $f_X(x \mid \theta_H)$.

The following generalization of the Neyman-Pearson fundamental result of Theorem 1 can be used to obtain the structure of an LO test:

Theorem 2: Let $g(x)$ and $h_1(x), h_2(x), ..., h_m(x)$ be real-valued and integrable functions defined on R^n. Let an integrable function $\delta(x)$ on R^n have the characteristics

$$\delta(x) = \begin{cases} 1 & , \quad g(x) > \sum_{i=1}^{m} t_i h_i(x) \\ r(x) & , \quad g(x) = \sum_{i=1}^{m} t_i h_i(x) \\ 0 & , \quad g(x) < \sum_{i=1}^{m} t_i h_i(x) \end{cases} \qquad (1\text{-}11)$$

for a set of constants $t_i \geq 0$, $i=1,2,...,m$, and where $0 \leq r(x) \leq 1$. Define, for $i=1,2,...,m$, the quantities

$$\alpha_i = \int_{R^n} \delta(x) h_i(x) \, dx \qquad (1\text{-}12)$$

Then from within the class of all test functions satisfying the m constraints (1-12), the function $\delta(x)$ defined by (1-11) maximizes $\int_{R^n} \delta(x) g(x) \, dx$.

A more complete version of the above theorem, and its proof, may be found in [Lehmann, 1959, Ch. 3]; Ferguson [1967, Ch. 5] also discusses the use of this result.

To use the above result in finding an LO test for $\theta = \theta_0$ against $\theta > \theta_0$ defining Θ_H and Θ_K in (1-1) and (1-2), respectively, let us write (1-9) explicitly as

$$p'(\theta_0 \mid \delta) = \frac{d}{d\theta} \int_{R^n} \delta(x) f_X(x \mid \theta) \, dx \bigg|_{\theta = \theta_0}$$

$$= \int_{R^n} \delta(x) \frac{d}{d\theta} f_X(x \mid \theta) \bigg|_{\theta = \theta_0} dx \qquad (1\text{-}13)$$

assuming that our pdf's are such as to allow the interchange of the order in which limits and integration operations are performed. Taking $m=1$ and identifying $h_1(\mathbf{x})$ with $f_\mathbf{X}(\mathbf{x}\mid\theta_0)$ and $g(\mathbf{x})$ with $\left.\dfrac{d}{d\theta}f_\mathbf{X}(\mathbf{x}\mid\theta)\right|_{\theta=\theta_0}$ in Theorem 2, we are led to the locally optimum test which accepts the alternative $K:\theta>\theta_0$ when

$$\frac{\left.\dfrac{d}{d\theta}f_\mathbf{X}(\mathbf{x}\mid\theta)\right|_{\theta=\theta_0}}{f_\mathbf{X}(\mathbf{x}\mid\theta_0)}>t \tag{1-14}$$

where t is the test threshold value which results in a size-α test satisfying

$$E\{\delta(\mathbf{X})\mid H:\theta=\theta_0\}=\alpha \tag{1-15}$$

The test of (1-14) may also be expressed as one accepting the alternative when

$$\left.\frac{d}{d\theta}\ln\{f_\mathbf{X}(\mathbf{x}\mid\theta)\}\right|_{\theta=\theta_0}>t \tag{1-16}$$

Theorem 2 may also be used to obtain tests maximizing the *second* derivative $p''(\theta_0\mid\delta)$ at $\theta=\theta_0$. This would be appropriate to attempt if it so happens that $p'(\theta_0\mid\delta)=0$ for all size-α tests for a given problem. The condition $p'(\theta_0\mid\delta)=0$ will occur if $\left.\dfrac{d}{d\theta}f_\mathbf{X}(\mathbf{x}\mid\theta)\right|_{\theta=\theta_0}$ is zero, assuming the requisite regularity conditions mentioned above. In this case Theorem 2 can be applied to obtain the locally optimum test accepting the alternative hypothesis $K:\theta>\theta_0$ when

$$\frac{\left.\dfrac{d^2}{d\theta^2}f_\mathbf{X}(\mathbf{x}\mid\theta)\right|_{\theta=\theta_0}}{f_\mathbf{X}(\mathbf{x}\mid\theta_0)}>t \tag{1-17}$$

One type of problem for which Theorem 2 is useful in characterizing locally optimum tests is that of testing $\theta=\theta_0$ against the *two-sided* alternative hypothesis $\theta\neq\theta_0$. We have previously mentioned that one can impose the condition of *unbiasedness* on the allowable tests for a problem. Unbiasedness of a size-α test for the hypotheses H and K of (1-1) and (1-2) means that the test satisfies

$$p(\theta \mid \delta) \leq \alpha, \text{ all } \theta \in \Theta_H \qquad (1\text{-}18)$$

$$p(\theta \mid \delta) \geq \alpha, \text{ all } \theta \in \Theta_K \qquad (1\text{-}19)$$

so that the detection probability for any $\theta_K \in \Theta_K$ is never less than the size α. For the two-sided alternative hypothesis $\theta \neq \theta_0$, suppose the pdf's $f_X(x \mid \theta)$ are sufficiently regular so that the power functions of all tests are twice continuously differentiable at $\theta = \theta_0$. Then it follows that for any *unbiased* size-α test we will have $p(\theta_0 \mid \delta) = \alpha$ and $p'(\theta_0 \mid \delta) = 0$. Thus, the test function of a locally optimum unbiased test can be characterized by using these two constraints and maximizing $p''(\theta_0 \mid \delta)$ in Theorem 2. Another interpretation of the above approach for the two-sided alternative hypothesis is that the quantity $\omega = (\theta - \theta_0)^2$ may then be used as a measure of the distance of any alternative hypothesis from the null hypothesis $\theta = \theta_0$. We have

$$\left. \frac{d}{d\omega} p(\theta \mid \delta) \right|_{\omega = 0} = \frac{1}{2(\theta - \theta_0)} \left. \frac{d}{d\theta} p(\theta \mid \delta) \right|_{\theta = \theta_0}$$

$$= \frac{1}{2} p''(\theta_0 \mid \delta) \qquad (1\text{-}20)$$

if $p'(\theta_0 \mid \delta)$ is zero, for sufficiently regular pdf's $f_X(x \mid \theta)$. Thus if $p'(\theta_0 \mid \delta)$ is zero for a class of size-α tests, then maximization of $p''(\theta_0 \mid \delta)$ leads to a test which is locally optimum within that class.

1.5 Bayes Tests

In general statistical decision theory which can treat estimation and hypothesis-testing problems within a single framework, there are four fundamental entities. These are (a) the observation space, which in our case is $\mathbb{R}^n$; (b) the set Θ of values of θ which parameterizes the possible distributions of the observations; (c) the set of all actions a which may be taken, the action space A; and (d) the loss function $l(\theta, a)$, a real-valued function measuring the loss suffered due to an action a when $\theta \in \Theta$ is the parameter value. In a binary hypothesis-testing problem the action space A will have only two possible actions, a_H and a_K, which, respectively, represent acceptance of the hypotheses H and K; and as a reasonable choice of loss function we can take

$$l(\theta, a) = c_{LJ}, \ \theta \in \Theta_J, \text{ and } a = a_L \qquad (1\text{-}21)$$

where $J, L = H$ or K, the c_{LJ} are non-negative, and $c_{JJ} = 0$. What is sought is a decision rule $d(\mathbf{x})$ taking on values in A, which specifies the action to be taken when an observation $\mathbf{x}$ has been made. More generally we can permit randomized decisions $\delta(\mathbf{x})$ which for each $\mathbf{x}$ specify a probability distribution over A.

The performance of any decision rule $d(\mathbf{x})$ can be characterized by the average loss that is incurred in using it; this is the *risk function*

$$R(\theta, d) = E\{l(\theta, d(\mathbf{x})) \mid \theta\}$$
$$= \int_{R^n} l(\theta, d(\mathbf{x})) f_X(\mathbf{x} \mid \theta) \, d\mathbf{x} \qquad (1\text{-}22)$$

The risk function for any given decision rule is nonetheless a function of θ, so that a comparison of the performances of different decision rules over a set of values of θ is not quite straightforward. A single real number serving as a figure of merit is assigned to a decision rule in Bayesian decision theory; to do this there is assumed to be available information leading to an *a priori* characterization of a probability distribution over Θ. We will denote the corresponding pdf as $\pi(\theta)$, and obtain the *Bayes risk* for a given prior density $\pi(\theta)$ and a decision rule $d(\mathbf{x})$ as

$$r(\pi, d) = E\{R(\theta, d)\}$$
$$= \int_{\Theta} R(\theta, d) \pi(\theta) \, d\theta \qquad (1\text{-}23)$$

In the binary hypothesis-testing problem of deciding between Θ_H and Θ_K for θ in $f_X(\mathbf{x} \mid \theta)$ [Equations (1-1) and (1-2)], the prior pdf $\pi(\theta)$ may be obtained as

$$\pi(\theta) = \pi_H \pi(\theta \mid H) + \pi_K \pi(\theta \mid K) \qquad (1\text{-}24)$$

where π_H and π_K are the respective *a priori* probabilities that H and K are true ($\pi_H + \pi_K = 1$), and $\pi(\theta \mid H)$ and $\pi(\theta \mid K)$ are the conditional *a priori* pdf's over Θ, conditioned, respectively, on H and K being true. For the loss function of (1-21) this gives

$$r(\pi, d) = \int_{\Theta} \int_{R^n} l(\theta, d(\mathbf{x})) f_X(\mathbf{x} \mid \theta) \pi(\theta) \, d\mathbf{x} \, d\theta$$
$$= \int_{R^n} \int_{\Theta} l(\theta, d(\mathbf{x})) f_X(\mathbf{x} \mid \theta) \left\{ \sum_{J=H,K} \pi_J \pi(\theta \mid J) \right\} d\theta \, d\mathbf{x}$$
$$(1\text{-}25)$$

For observations **x** for which $d(\mathbf{x}) = a_H$, the inner integral over Θ becomes

$$\pi_K c_{HK} \int_{\Theta_K} f_\mathbf{X}(\mathbf{x} \mid \theta)\pi(\theta \mid K) \, d\theta$$

$$= \pi_K c_{HK} f_\mathbf{X}(\mathbf{x} \mid K) \qquad (1\text{-}26)$$

where $f_\mathbf{X}(\mathbf{x} \mid K)$ is the conditional pdf of **X** given that the alternative hypothesis K is true. Similarly, for **x** such that $d(\mathbf{x}) = a_K$ the integral over Θ in (1-25) becomes $\pi_H c_{KH} f_\mathbf{X}(\mathbf{x} \mid H)$. Thus the Bayes rule minimizing $r(\pi, d)$ accepts the alternative hypothesis K when

$$\frac{f_\mathbf{X}(\mathbf{x} \mid K)}{f_\mathbf{X}(\mathbf{x} \mid H)} > \frac{\pi_H c_{KH}}{\pi_K c_{HK}} \qquad (1\text{-}27)$$

When the likelihood ratio on the left-hand side above is strictly less than the threshold value on the right-hand side, the null hypothesis H is accepted by the Bayes rule. When equality holds $d(\mathbf{x})$ may specify any choice between a_H and a_K.

Note that for a test implementing Bayes' rule the threshold value is completely specified once the prior probabilities π_H and π_K and costs c_{LJ} are fixed; the false-alarm probability does not enter as a consideration in setting the threshold. The development above assumed that $\pi(\theta \mid K)$ and $\pi(\theta \mid H)$ are available when the hypotheses are composite. As an alternative approach when Θ_K is composite, one can consider *locally optimum* Bayes' tests for which only the first few terms in a power series expansion of the likelihood ratio $f_\mathbf{X}(\mathbf{x} \mid \theta_K)/f_\mathbf{X}(\mathbf{x} \mid \theta_H)$ are used on the left-hand side of (1-27). Such an approach has been considered by Middleton [1966, 1984] and we will describe its application in a detection problem in Section 2.5 of the next chapter.

1.6 A Characterization of the Relative Performance of Tests

In this final section before we proceed to consider signal detection problems we will develop a relative-performance measure which is conveniently applied and which is at the same time of considerable value in obtaining simple quantitative comparisons of different tests or detectors. This measure is called the *asymptotic relative efficiency* and we will use it, together with a detector performance measure related to it called the *efficacy*, quite extensively in this book.

Let D_A and D_B be two detectors based on test statistics $T_A(\mathbf{X})$ and $T_B(\mathbf{X})$, respectively, so that the test function $\delta_A(\mathbf{X})$ of

D_A is defined by

$$\delta_A(\mathbf{x}) = \begin{cases} 1 & , \quad T_A(\mathbf{x}) > t_A \\ r_A & , \quad T_A(\mathbf{x}) = t_A \\ 0 & , \quad T_A(\mathbf{x}) < t_A \end{cases} \qquad (1\text{-}28)$$

with a similar definition for $\delta_B(\mathbf{x})$, in terms of respective thresholds t_A and t_B and randomization probabilities r_A and r_B. Suppose the thresholds t_A and t_B (and the randomization probabilities r_A and r_B, if necessary) are designed so that both detectors have the same size or false-alarm probability $p_f = \alpha$. To compare the relative detection performance or power of the two detectors one would have to obtain the power functions of the two detectors, that is, detection probabilities would have to be computed for all $\theta \in \Theta_K$. Furthermore, such a comparison would only be valid for one particular value of the size, α, so that power functions would have to be evaluated at all other values of the size which may be of interest. In addition, while it has not yet been made very explicit, such a comparison can be expected to depend on the number of observation components n used by the detectors when n is a design parameter; for example, the observation components X_i, $i = 1,2,...,n$, may be identically distributed outcomes of repeated independent observations, forming a random sample of size n, governed by a univariate pdf $f_X(x \mid \theta)$.

In any event one would still be faced with the problem of expressing in a succinct and useful way the result of such a performance comparison. Of particular interest would be a real-valued measure which could be taken as an index of the overall relative performance of two tests or detectors. It would be even more appealing if such a measure could be computed directly from general formulas which would not require any explicit computation of families of power functions. As a step toward the definition of such a single index of relative performance, let us discuss a measure which is called the *relative efficiency* of two tests.

1.6.1 Relative Efficiency

Suppose that the number of observation components n in $\mathbf{X}$ is a variable quantity, so that any given detector has a choice of observation-vector size n that it can operate on for some given hypothesis-testing problem, namely testing $\theta \in \Theta_H$ against $\theta \in \Theta_K$ for θ in $f_\mathbf{X}(\mathbf{x} \mid \theta)$. For example, the linear detector test statistic is $\sum_{i=1}^{n} X_i$, and such a detector may be designed to operate on any sample size n. The situation which generally holds is that increasing n leads to improvement in the performance of a detector; for a fixed design value of the size α, for example, the detection probability or power at any value of $\theta \in \Theta_K$ will increase with n. It is also generally true that there is a cost associated with the use of a larger number of observations, which in signal detection applications may mean a higher sampling rate, a longer observation time interval and therefore longer delay before a decision is made, or a heavier computational burden. Thus one measure of the relative performance of two tests or detectors in any given hypothesis-testing problem is obtained as a ratio of observation sizes required by the two procedures to attain a given level of performance.

Consider any specific alternative hypothesis defined by some θ_K in Θ_K, and let us consider detectors for $\theta \in \Theta_H$ against $\theta \in \Theta_K$ which are of size α and have detection probability $p_d = 1 - \beta$ for $\theta = \theta_K$, where the specified α and "miss" probability β are between 0 and 1. Let D_A and D_B be two detectors achieving this performance specification, based on respective observation vector lengths n_A and n_B. These observation vector lengths n_A and n_B are clearly functions of α, β and θ_K and are more explicitly expressed as $n_A(\alpha,\beta,\theta_K)$ and $n_B(\alpha,\beta,\theta_K)$, respectively.

The *relative efficiency* $RE_{A,B}$ of the detectors D_A and D_B is defined as the ratio

$$RE_{A,B} = \frac{n_B(\alpha,\beta,\theta_K)}{n_A(\alpha,\beta,\theta_K)} \tag{1-29}$$

This obviously depends on α, β and θ_K in general. In addition to this dependence of the value of $RE_{A,B}$ on the operating point (α,β,θ_K), the computation of this quantity requires knowledge of the probability distribution functions of $T_A(\mathbf{X})$ and $T_B(\mathbf{X})$ for $\theta = \theta_K$. In order to alleviate this requirement, which is often difficult to meet, one could look at the asymptotic case where both n_A and n_B become large, with the expectation that the limiting distributions of the test statistics become Gaussian. It turns out that under the proper asymptotic formulation of the problem the relative efficiency converges to a quantity which is independent of α and β and is much easier to evaluate in practice.

1.6.2 Asymptotic Relative Efficiency and Efficacy

For $\theta = \theta_K$ a fixed parameter in the alternative hypothesis observation pdf $f_{\mathbf{X}}(\mathbf{x} \mid \theta)$, consider the sequence of observation-vector lengths $n = 1,2,\ldots$. Almost without exception, tests of hypotheses used in applications such as signal detection have the property that, for fixed α, the power of the tests increase to a limiting value of unity (or $\beta \to 0$) for $n \to \infty$. Thus in seeking an asymptotic definition of relative efficiency this effect needs to be addressed. Let us first formalize this type of behavior by giving a definition of the property of *consistency* of tests.

The H and K of (1-1) and (1-2) were stated for the pdf of a vector of length n of observations. Consider now a *sequence* $\{\theta_n, n = 1,2,...\}$ of θ values in Θ_K, and the sequence of corresponding hypothesis-testing problems that we get for $n = 1,2,...$:

$$H_n : \mathbf{X} \text{ has pdf } f_\mathbf{X}(\mathbf{x} \mid \theta) \text{ with } \theta \in \Theta_H \qquad (1\text{-}30)$$

$$K_n : \mathbf{X} \text{ has pdf } f_\mathbf{X}(\mathbf{x} \mid \theta_n) \qquad (1\text{-}31)$$

Let $\{T_n(\mathbf{X}), n = 1,2,...\}$ be the *sequence* of test statistics, with $\mathbf{X} = (X_1, X_2, ..., X_n)$, of some detector type. This generally means that $T_n(\mathbf{X})$ for different n has a fixed functional form; for example, $T_n(\mathbf{X}) = \sum_{i=1}^{n} X_i$. Now the sequence of individual detectors D_n based on the $T_n(\mathbf{X})$ (and corresponding threshold and randomization probabilities t_n and r_n, respectively) is *asymptotically of size* α in testing $\theta \in \Theta_H$ if

$$\lim_{n \to \infty} \sup_{\theta \in \Theta_H} E\{\delta_n(\mathbf{X}) \mid \theta\} = \alpha \qquad (1\text{-}32)$$

Here $\delta_n(\mathbf{X})$ is the test function obtained using the statistic $T_n(\mathbf{X})$. Such an asymptotically size-α sequence $\{D_n, n = 1,2,...\}$ of detectors is said to be *consistent* for the sequence of alternatives $\{\theta_n, n = 1,2,...\}$ if

$$\lim_{n \to \infty} p(\theta_n \mid \delta_n) = \lim_{n \to \infty} p_d(\theta_n \mid D_n)$$
$$= 1 \qquad (1\text{-}33)$$

We have remarked above that if θ_n is fixed to be some θ_K in Θ_K, sequences of most of the types of tests of interest to us will be consistent. In order to consider asymptotic situations where observation lengths n become large in relative efficiency considerations, one approach therefore is to define sequences of alternatives parameterized by the θ_n for which the limiting power is not unity but is a specified value $1-\beta$ (with $0 < \beta < 1$). This clearly implies that the θ_n should approach a null-hypothesis parameter value as $n \to \infty$. We have stated earlier that the local case in hypothesis testing, where null- and alternative-hypothesis distributions of the observations approach one another, is of considerable interest to us since in many signal detection problems the weak-signal case is of particular concern. Thus we should expect that such an asymptotic formulation of the hypothesis-testing problem will be quite useful, corresponding to a situation of a weak signal and a long observation record in signal detection.

Let us now focus on hypothesis-testing problems where the parameter θ is *real-valued*, and further constrain the null hypothesis to be *simple*. In particular, without loss of generality, let $\theta = 0$ be taken to describe the simple null hypothesis. Furthermore, let us consider the one-sided alternative hypothesis described by $\theta > 0$. We are now ready to show that an asymptotic version of the relative efficiency may be defined and computed quite readily for two tests, under some regularity conditions which are frequently satisfied.

Let $\{\theta_l, l=1,2,...\}$ be a sequence of alternative-hypothesis parameters converging to $\theta = 0$, and let $\{n_{A,l}\}$ and $\{n_{B,l}\}$ be two corresponding sequences of observation-vector lengths for two detector sequences $\{D_A^l\}$ and $\{D_B^l\}$, respectively, so that D_A^l uses $n_{A,l}$ observations X_i in testing $\theta = 0$ against $\theta = \theta_l$, and similarly, D_B^l uses an observation-vector length of $n_{B,l}$.

Definition: Let $\{D_A^l\}$ and $\{D_B^l\}$ be of asymptotic size α for $\theta = 0$, and let their limiting power values for $l \to \infty$ exist and be equal for the alternatives $\{\theta_l\}$,

$$\lim_{l \to \infty} p_d(\theta_l \mid D_A^l) = \lim_{l \to \infty} p_d(\theta_l \mid D_B^l)$$
$$= 1 - \beta \qquad (1\text{-}34)$$

with $0 < \beta < 1$. Then the *asymptotic relative efficiency* (ARE) of $\{D_A^l\}$ with respect to $\{D_B^l\}$ is defined by

$$ARE_{A,B} = \lim_{l \to \infty} \frac{n_{B,l}}{n_{A,l}} \qquad (1\text{-}35)$$

provided that the limiting value in (1-35) is independent of α and β and of the particular sequences $\{\theta_l\}$, $\{n_{A,l}\}$, and $\{n_{B,l}\}$ satisfying (1-34).

The above definition of the ARE is clearly an asymptotic version of the relative efficiency. The main feature of this asymptotic definition is that we have considered a sequence of alternative-hypothesis parameter values which approach the null-hypothesis value $\theta = 0$, and corresponding observation-vector lengths which grow in such a way that the *asymptotic* powers become some value $1 - \beta$ between zero and unity. Note that we also require the ARE to be independent of the asymptotic size α of the detectors and of β. As we have remarked before, the detectors D_A^l (or D_B^l) are for different l members of a particular type of procedure having a common functional dependence of its test statistic on $\mathbf{X}$, and we often speak of such a whole sequence of detectors as a detector D_A (or D_B).

Regularity Conditions

Consider a sequence of detectors $\{D_n\}$ based on a corresponding sequence of test statistics $\{T_n(\mathbf{X})\}$ with threshold sequence $\{t_n\}$. Let the detector sequence be of asymptotic size α, and let $\mu_n(\theta)$ and $\sigma_n^2(\theta)$ be the mean and variance, respectively, of $T_n(\mathbf{X})$ for $\theta \geq 0$. The following regularity conditions make computation of the ARE quite simple in many cases:

(i) $\left. \dfrac{d}{d\theta} \mu_n(\theta) \right|_{\theta=0} = \mu_n'(0)$ exists and is positive;

(ii) $$\lim_{n \to \infty} \frac{\mu_n'(0)}{\sqrt{n}\ \sigma_n(0)} = c > 0; \tag{1-36}$$

For $\theta_n = \gamma/\sqrt{n}$, with $\gamma \geq 0$,

(iii) $$\lim_{n \to \infty} \frac{\mu_n'(\theta_n)}{\mu_n'(0)} = 1 \tag{1-37}$$

and

$$\lim_{n \to \infty} \frac{\sigma_n(\theta_n)}{\sigma_n(0)} = 1; \tag{1-38}$$

(iv) $[T_n(\mathbf{X}) - \mu_n(\theta_n)]/\sigma_n(\theta_n)$ has asymptotically the standard normal distribution.

If these regularity conditions are satisfied by two given detectors, we can show that their ARE can be computed quite easily. Notice that for conditions (iii) and (iv) above a particular rate of convergence of θ_n to zero is used, so that in making ARE evaluations using these conditions, we are assuming this type of convergence of θ_n to zero. We should also remark that in many cases which will be of interest to us condition (iv) may be shown to hold; this is usually true in particular when $\mathbf{X} = (X_1, X_2, \ldots, X_n)$ is a vector of independent components and the test statistic $T_n(\mathbf{X})$ is the log of the likelihood ratio. The regularity conditions (i) - (iii) also hold for many test statistics of interest. The D_n being of asymptotic size α, it follows easily from condition (iv) that $[t_n - \mu_n(0)]/\sigma_n(0)$ converges to z_d where

$$\alpha = 1 - \Phi(z_d) \tag{1-39}$$

Φ being the standard normal distribution function. We also get from the regularity conditions

$$\lim_{n \to \infty} p_d(\theta_n \mid D_n) = \lim_{n \to \infty} p\{T_n(\mathbf{X}) > t_n \mid \theta_n\}$$
$$= 1 - \Phi(z_d - \gamma c) \qquad (1\text{-}40)$$

where c was defined in (1-36) and $\theta_n = \gamma/\sqrt{n}$. The quantity c^2 is called the *efficacy*[1] ξ of the detector sequence $\{D_n\}$, which is therefore

$$\xi = \lim_{n \to \infty} \frac{\left[\frac{d}{d\theta} E\{T_n(\mathbf{X}) \mid \theta\}\Big|_{\theta=0}\right]^2}{n \, V\{T_n(\mathbf{X}) \mid \theta\}\Big|_{\theta=0}} \qquad (1\text{-}41)$$

From these results we get the following theorem:

Theorem 3: Let D_A and D_B be asymptotic size-α detector sequences whose test statistics satisfy the regularity conditions (i) - (iv). Then the $ARE_{A,B}$ of D_A relative to D_B is

$$ARE_{A,B} = \frac{\xi_A}{\xi_B} \qquad (1\text{-}42)$$

where ξ_A and ξ_B are the efficacies of D_A and D_B, respectively.

To prove this result, note that if $\gamma = \gamma_A > 0$ and $\gamma = \gamma_B > 0$ define the sequences of alternatives $\theta_n = \gamma/\sqrt{n}$ for D_A and D_B, respectively, then the detector sequences $\{D_{A,n}\}$ and $\{D_{B,n}\}$ have the same asymptotic power if

$$\gamma_A \, \xi_A^{1/2} = \gamma_B \, \xi_B^{1/2} \qquad (1\text{-}43)$$

On the other hand, consider two subsequences $\{D_A^l\}$ and $\{D_B^l\}$ from $\{D_{A,n}\}$ and $\{D_{B,n}\}$, respectively, such that asymptotically, for $l \to \infty$,

$$\frac{\gamma_A}{\sqrt{n_{A,l}}} = \frac{\gamma_B}{\sqrt{n_{B,l}}} \qquad (1\text{-}44)$$

where γ_A and γ_B satisfy (1-43). Then for $\theta_l = \gamma_A/\sqrt{n_{A,l}}$ both $\{D_A^l\}$ and $\{D_B^l\}$ have the same limiting power as $l \to \infty$, given by (1-40) with $\gamma c = \gamma_A \, \xi_A^{1/2} = \gamma_B \, \xi_B^{1/2}$. From (1-43) and (1-44), and the definition (1-35), the result of Theorem 3 follows easily.

---

[1] The quantity c itself is usually called the efficacy in statistics.

Thus we find that, subject to the regularity conditions (i) - (iv), the ARE of two detectors can be computed quite simply as a ratio of their efficacies; the efficacy of a detector is obtained from (1-41), which can be seen to require derivation of the means and variances only of the test statistics. We will be basing most of our performance comparisons in this book on the ARE, so that in the following chapters we will apply Theorem 3 and (1-41) many times. In the next chapter we will enter into a further discussion of the ARE for the special case of detection of known signals in additive noise, which will serve to illustrate more explicitly the ideas we have developed here.

Extended Regularity Conditions

As a generalization of the above results, we can consider the following extended versions of the conditions (i) - (iv):

(i)′ $\left. \dfrac{d^i}{d\theta^i} \mu_n(\theta) \right|_{\theta=0} = \mu_n^{(i)}(0)$

$$= 0, \quad i=1,2,\ldots,m-1,$$

and

$$\mu_n^{(m)}(0) > 0$$

for some integer m;

(ii)′ $\displaystyle\lim_{n \to \infty} \dfrac{\mu_n^{(m)}(0)}{n^{m\delta}\sigma_n(0)} = c > 0$ \hfill (1-45)

for some $\delta > 0$;

For $\theta_n = \gamma/n^\delta$, with $\gamma \geq 0$,

(iii)′ $\displaystyle\lim_{n \to \infty} \dfrac{\mu_n^{(m)}(\theta_n)}{\mu_n^{(m)}(0)} = 1$ \hfill (1-46)

and

$\displaystyle\lim_{n \to \infty} \dfrac{\sigma_n(\theta_n)}{\sigma_n(0)} = 1;$ \hfill (1-47) (iv)' $[T_n(\mathbf{X}) - \mu_n(\theta_n)]/\sigma_n(\theta_n)$ has asymptotically the standard normal distribution.

If we now define the efficacy by

$$\xi = \lim_{n \to \infty} \frac{1}{n} \left[\frac{\mu_n^{(m)}(0)}{\sigma_n(0)} \right]^{1/m\delta} \tag{1-48}$$

the ARE of two detectors satisfying the regularity conditions (i)' - (iv)' for a *common* δ and the same value of m can again be obtained as a ratio of their efficacies. Note that $m=1$ and $\delta=1/2$ is the case we considered first; it is the most commonly occurring case.

While we have focused on the case of alternatives $\theta > 0$, it should be quite clear that for alternatives $\theta < 0$ exactly the same results hold, which can be established by reparameterizing with $\theta_c = -\theta$.

The extended results above also allow consideration of certain tests for the two-sided alternatives $\theta \neq 0$. For detailed discussions of the asymptotic comparisons of tests the reader may consult the book by Kendall and Stuart [1967]. The original investigation of this type of asymptotic performance evaluation was made by Pitman [1948] and generalized by Noether [1955].

The Multivariate Case

In the discussion above of the efficacy of a test and of the ARE of two tests we required that the distributions of the test statistics be asymptotically normal. A careful examination of the above development reveals, however, that it is not the asymptotic *normality* of the test statistics which is the key requirement. What is needed to obtain the ARE of two tests is that their test statistics have the *same* functional form for their asymptotic distributions, so that the asymptotic powers of the two tests based on them can be made the same by proper choice of the sample sizes. Later in this book we will consider test statistics which may be expressed as quadratic forms having asymptotically the χ^2 distribution. Let us therefore extend our definition of the efficacy of a test to allow us to treat such cases.

Consider a sequence $\{\mathbf{T}_n(\mathbf{X})\}$ of *multivariate* statistics with

$$\mathbf{T}_n(\mathbf{X}) = [T_{1n}(\mathbf{X}), T_{2n}(\mathbf{X}), ..., T_{Ln}(\mathbf{X})] \tag{1-49}$$

Let the means and variances of the components for $\theta \geq 0$ be

$$E\{T_{jn}(\mathbf{X}) \mid \theta\} = \mu_{jn}(\theta), \quad j = 1,2,...,L \tag{1-50}$$

$$V\{T_{jn}(\mathbf{X}) \mid \theta\} = \sigma_{jn}^2(\theta), \quad j = 1,2,...,L \tag{1-51}$$

and let the covariances for $\theta \geq 0$ be

$$\text{COV}\{T_{kn}(\mathbf{X}) \, T_{jn}(\mathbf{X}) \mid \theta\} = \rho_{kjn}(\theta) \, \sigma_{kn}(\theta) \, \sigma_{jn}(\theta) \tag{1-52}$$

so that $\rho_{kjn}(\theta)$ is the normalized covariance or coefficient of correlation. We now impose the multivariate versions of the extended regularity conditions given above for the scalar case:

(i)'' $\left. \dfrac{d^i}{d\theta^i} \mu_{jn}(\theta) \right|_{\theta=0} = \mu_{jn}^{(i)}(0)$

$\qquad\qquad\qquad = 0, \quad i = 1,2,...,m-1$

and

$\mu_{jn}^{(m)}(0) > 0$

for some integer m, for $j = 1,2,...,L$;

(ii)'' $\displaystyle\lim_{n \to \infty} \dfrac{\mu_{jn}^{(m)}(0)}{n^{m\delta} \sigma_{jn}(0)} = c_j > 0 \tag{1-53}$

for some $\delta > 0$, for $j = 1,2,...,L$;

For $\theta_n = \gamma/n^\delta$, with $\gamma \geq 0$,

(iii)'' $\displaystyle\lim_{n \to \infty} \dfrac{\mu_{jn}^{(m)}(\theta_n)}{\mu_{jn}^{(m)}(0)} = 1 \tag{1-54}$

$\displaystyle\lim_{n \to \infty} \dfrac{\sigma_{jn}(\theta_n)}{\sigma_{jn}(0)} = 1 \tag{1-55}$

and

$\displaystyle\lim_{n \to \infty} \dfrac{\rho_{kjn}(\theta_n)}{\rho_{kjn}(0)} = 1 \tag{1-56}$

for $k,j = 1,2,...,L$;

(iv)'' the $[T_{jn}(\mathbf{X}) - \mu_{jn}(\theta_n)]/\sigma_{jn}(\theta_n)$, for $j = 1,2,...,L$, have asymptotically a multivariate normal distribution.

Let us define the normalized versions $Q_{jn}(\mathbf{X},\theta)$ of the components $T_{jn}(\mathbf{X})$ of the multivariate statistics as

$$Q_{jn}(\mathbf{X},\theta) = \frac{T_{jn}(\mathbf{X}) - \mu_{jn}(\theta)}{\sigma_{jn}(\theta)}, \quad j = 1,2,...,L \qquad (1\text{-}57)$$

and let $\mathbf{Q}_n(\mathbf{X},\theta)$ be the row vector of components $Q_{jn}(\mathbf{X},\theta)$. Now a real-valued statistic of particular interest is that defined by the quadratic form

$$U_n(\mathbf{X}) = \mathbf{Q}_n(\mathbf{X},0) \, \mathbf{R}_n^{-1}(0) \, \mathbf{Q}_n(\mathbf{X},0)^T \qquad (1\text{-}58)$$

where the matrix $\mathbf{R}_n(\theta)$ is the normalized $L \times L$ covariance matrix of elements $\rho_{kjn}(\theta)$. Under the null hypothesis $\theta = 0$ this statistic has asymptotically a χ^2 distribution with L degrees of freedom. To obtain its limiting distribution under the alternatives $\theta_n = \gamma/n^\delta$, note that $Q_{jn}(\mathbf{X},0)$ can be expressed as

$$Q_{jn}(\mathbf{X},0) = Q_{jn}(\mathbf{X},\theta_n) \frac{\sigma_{jn}(\theta_n)}{\sigma_{jn}(0)} + \frac{\mu_{jn}(\theta_n) - \mu_{jn}(0)}{\sigma_{jn}(0)} \qquad (1\text{-}59)$$

so that it is asymptotically equivalent to $Q_{jn}(\mathbf{X},\theta_n) + (\gamma^m/m!)c_j$, where c_j was defined by (1-53) in (ii)''. Using condition (iv)'' we conclude finally that $U_n(\mathbf{X})$ has asymptotically a *non-central* χ^2 distribution with L degrees of freedom under the sequence of alternative hypotheses defined by $\theta_n = \gamma/n^\delta$. The non-centrality parameter in this distribution is

$$\Delta = \frac{\gamma^{2m}}{(m!)^2} \mathbf{c} \, \mathbf{R}^{-1} \, \mathbf{c}^T \qquad (1\text{-}60)$$

where $\mathbf{c} = (c_1, c_2, \ldots, c_L)$ and $\mathbf{R} = \lim_{n \to \infty} \mathbf{R}_n(0)$.

Suppose we have two tests based on L-variate quantities $\mathbf{T}_{A,n}(\mathbf{X})$ and $\mathbf{T}_{B,n}(\mathbf{X})$ satisfying the multivariate extended regularity conditions (i)'' - (iv)''. Let their respective vectors $\mathbf{c}$ of individual efficacy components be denoted as $\mathbf{c}_A$ and $\mathbf{c}_B$, and let their respective limiting normalized covariance matrices be $\mathbf{R}_A$ and $\mathbf{R}_B$. Assume further that the conditions (i)'' - (iv)'' are satisfied for the *same* values of m and δ by both sequences of statistics, and consider the quadratic form test statistics defined by (1-58) for each of the two tests. It follows exactly as in the scalar cases treated above that for the same sequence of alternatives $\theta_n = \gamma/n^\delta$, the $ARE_{A,B}$ of the two tests is

$$ARE_{A,B} = \left[\frac{c_A R_A^{-1} c_A^T}{c_B R_B^{-1} c_B^T} \right]^{1/(2m\,\delta)} \tag{1-61}$$

We may therefore define the *generalized efficacy* of a test based on a quadratic form statistic satisfying our assumptions (i)'' - (iv)'' as

$$\xi = [c\, R^{-1}\, c^T]^{1/(2m\,\delta)} \tag{1-62}$$

Notice that this reduces to the efficacy defined by (1-48) for the scalar case $L = 1$, and further to the efficacy defined by (1-41) when $m = 1$ and $\delta = 1/2$.

PROBLEMS

Problem 1.1

$X_1, X_2, ..., X_n$ are independent real-valued observations governed by the pdf

$$f(x \mid \theta) = \theta\, x^{\theta-1},\ 0 \le x \le 1$$

where the parameter θ has a value in $[1,\infty)$.

(a) Find the form of the best test for $H:\ \theta = \theta_0$ versus $K:\ \theta = \theta_1$. Show that the test can be interpreted as a comparison of the sum of a nonlinear function of each observation component to a threshold.

(b) Determine explicitly the best test of size $\alpha = 0.1$ for $\theta = 2$ versus $\theta = 3$. Is the test uniformly most powerful for testing $\theta = 2$ versus $\theta > 2$?

Problem 1.2

An observation X has pdf

$$f_X(x \mid \theta) = \frac{a}{2}\, e^{-a|x-\theta|},\ -\infty < x < \infty$$

where $a > 0$ is known. Consider size α tests, $0 < \alpha < 1$, for $\theta = 0$ versus $\theta > 0$ based on the single observation X.

(a) Show that a test which rejects $\theta = 0$ when $X > t$ is uniformly most powerful for $\theta > 0$.

(b) Show that a test which rejects $\theta = 0$ with probability γ when $X > 0$ is also a locally optimum test for $\theta > 0$.

(c) Sketch the power functions of these two tests. Verify that they have the same slope at $\theta = 0$.

Problem 1.3

$\mathbf{X} = (X_1, X_2, ..., X_n)$ is a vector of independent and identically distributed observations, each governed by the pdf

$$f(x \mid \theta, a) = ae^{-a(x-\theta)}, \quad \theta < x < \infty$$

where θ and a are real-valued parameters. We want to test the simple null hypothesis $H: \theta = \theta_0, a = a_0$ against the simple alternative hypothesis $K: \theta = \theta_1 \leq \theta_0, a = a_1 > a_0$.

(a) Find in its simplest form the best test for H versus K. Let the alternative hypothesis K_1 be defined by $K_1: \theta \leq \theta_0, a > a_0$. Is your test uniformly most powerful for K_1?

(b) Let K_2 be the alternative hypothesis $\theta = \theta_0, a > a_0$. Is the above test uniformly most powerful for K_2?

(c) Obtain the generalized likelihood ratio test for H versus K_2, and compare it with your test found in (a).

Problem 1.4

$\mathbf{X} = (X_1, X_2, ..., X_n)$ is a vector of independent and identically distributed components, each being Gaussian with mean μ and variance σ^2. For testing $H: \sigma^2 = 1$, μ unspecified, versus $K: \sigma^2 \neq 1$, μ unspecified, obtain the form of the *generalized likelihood ratio test*. Show that the test reduces to deciding if a particular test statistic falls inside some interval. Explain how the end points of the interval can be determined to obtain a specified value for the size α.

Problem 1.5

$\mathbf{X} = (X_1, X_2, \cdots, X_n)$ is a vector of independent and identically distributed Gaussian components, each with mean θ and variance 1. Let $\bar{X}$ be the sample mean $\sum_{i=1}^{n} X_i / n$. For testing $\theta = 0$ versus $\theta \neq 0$, show that the test which rejects $\theta = 0$ when $|\bar{X}|$ exceeds a threshold is uniformly most powerful unbiased of its size. (Use Theorem 2.)

Problem 1.6

Given n independent and identically distributed observations X_i governed by a unit-variance Gaussian pdf with mean θ, a test for $\theta = 0$ versus $\theta > 0$ may be based on the test statistic $\sum_{i=1}^{n} X_i$. Another possible test statistic for this problem is $\sum_{i=1}^{n} X_i^2$. Obtain the efficacies for these two tests, for suitable sequences of alternatives $\{\theta_n\}$ converging to 0, verifying that the regularity conditions (i) - (iv) or (i)' - (iv)' are satisfied. What can you say about the ARE of these two tests?

Problem 1.7

For the Cauchy density function

$$f(x \mid \theta) = \frac{1}{\pi} \frac{1}{1 + (x - \theta)^2}$$

with location parameter θ, a test for $\theta = 0$ versus $\theta > 0$ may be based on the *median* of n independent observations. Using results on the distributions of order statistics, obtain the efficacy of this test.

Problem 1.8

Independent observations $X_1, X_2, \cdots, X_n$ are governed by a common pdf f which is unimodal and symmetric about its location value θ. The pdf f has a finite variance σ^2. Show that the ARE of the test for $\theta = 0$ based on the sample *median* relative to that based on the sample *mean* is $4\sigma^2 f^2(0)$. This result is independent of the scale parameter σ. Obtain its minimum value, and the corresponding unimodal symmetric pdf for which this minimum value of the ARE is achieved.

Chapter 2

DETECTION OF KNOWN SIGNALS IN ADDITIVE NOISE

2.1 Introduction

In this chapter we will begin our description of signal detection schemes for non-Gaussian noise by considering one of the simplest to formulate and most commonly encountered of all detection problems. This is the detection of a real-valued deterministic signal which is completely known in a background of additive noise. We shall be concerned exclusively with discrete-time detection problems in this book. In many applications discrete-time observations are obtained by sampling at a uniform rate some underlying continuous-time observation process. In the next section we will give the example of detection of a pulse train in additive noise to illustrate how a discrete-time detection problem may arise in a different manner from an original problem formulation in the continuous-time domain.

To obtain canonical results for the detector structures we will focus on the case of weak-signal detection when a large number of observation values are available, so that local expansions and asymptotic approximations can be used. We will then explain how the asymptotic performances of different detectors may be compared. In addition, we will discuss the value of such asymptotic performance comparison results for comparisons of performance in non-asymptotic cases of finite-length observation sequences and non-vanishing signal strengths.

Most of this development is for the detection problem in which the observations represent either noise only or a signal with additive noise. A common and important example of such a detection requirement is to be found in radar systems. In the communication context this detection requirement occurs in the case of on-off signaling. Furthermore, our development will be based primarily on the Neyman-Pearson approach, in which the performance criterion is based on the probability of correct detection for a fixed value of the probability of false detection or false alarm. In the penultimate section, however, we briefly discuss the case of binary signaling and the weak-signal version of the Bayes detector.

2.2 The Observation Model

For the known-signal-in-additive-noise detection problem we may describe our observation vector $\mathbf{X} = (X_1, X_2, ..., X_n)$ of real-valued components X_i by

$$X_i = \theta s_i + W_i, \quad i = 1,2,...,n \qquad (2\text{-}1)$$

Here the W_i are random noise components which are always present, and the s_i are the known values of the signal components. The signal amplitude θ is either zero (the observations contain no signal) or positive, and the detection problem is to decide on the basis of **X** whether we have $\theta = 0$ or we have $\theta > 0$.

2.2.1 Detection of a Pulse Train

As we have mentioned above, the X_i may be samples obtained from some continuous-time observation process. One such example occurs in the detection of a pulse train with a known pattern of amplitudes for the otherwise identical pulses in the train. For instance, the signal to be detected may be the bipolar pulse train depicted in Figure 2.1. In this case the X_i could be samples taken at the peak positions of the pulses (if present) from a noisy observed waveform. Alternatively, the X_i could be the outputs at specific time instants of some pre-detection processor, such as a filter matched to the basic pulse shape. It will be useful to consider a little more explicitly this mechanism for generating the observations X_i, before we continue with the model of (2-1).

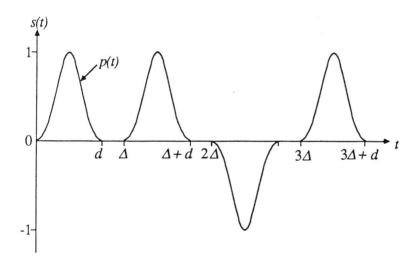

Figure 2.1 Bipolar Pulse Train

Let $p(t)$ be the basic pulse shape in Figure 2.1, defined to be zero outside the interval $(0, \Delta)$. The train of n such non-overlapping pulses may be described by

$$s(t) = \sum_{i=1}^{n} e_i \, p(t - [i-1]\Delta) \qquad (2\text{-}2)$$

where e_i is the known amplitude of the i-th pulse. In Figure 2.1 the amplitudes e_i are all 1 or -1. The received continuous-time observation process $X(t)$ may be expressed as

$$X(t) = \theta s(t) + W(t), \quad T_0 \leq t \leq T_1 \qquad (2\text{-}3)$$

Now $W(t)$ is an additive continuous-time random noise process. In (2-3) the observation interval $[T_0, T_1]$ includes the signal interval $[0, n\Delta]$.

Let the noise process $W(t)$ have zero mean and let it be wide-sense stationary with power spectral density $\Phi_W(\omega)$, and let $P(\omega)$ be the Fourier transform of the pulse $p(t)$. Suppose we wanted to find that *linear* processor which maximizes its output signal-to-noise ratio (SNR) at time $t = n\Delta$, the end of the pulse train, when the input is $X(t)$. The solution is the well-known *matched filter*, which is specified to have frequency response

$$\tilde{H}_M(\omega) = \frac{\sum_{i=1}^{n} e_i P^*(\omega) e^{j\omega(i-1)\Delta}}{\Phi_W(\omega)} \, e^{-j\omega n \Delta} \qquad (2\text{-}4)$$

(Strictly speaking, this is the correct solution for long observation intervals.) The numerator above (without the delay term $e^{-j\omega n \Delta}$) is the Fourier transform of $s(-t)$. From the above we find that

$$\tilde{H}_M(\omega) = \frac{P^*(\omega)}{\Phi_W(\omega)} \, e^{-j\omega \Delta} \sum_{i=1}^{n} e_i \, e^{-j\omega(n-i)\Delta}$$

$$= H_M(\omega) \sum_{i=1}^{n} e_i \, e^{-j\omega(n-i)\Delta} \qquad (2\text{-}5)$$

where $H_M(\omega)$ is the frequency response of the filter matched for a *single pulse* $p(t)$, maximizing output SNR for this case at time $t = \Delta$.

Thus the maximum output SNR linear filter can be given the system interpretation of Figure 2.2. Here the output of the *single-pulse matched filter* is sampled at times $t = i\Delta$, $i = 1,2,\ldots,n$, the i-th sampled value X_i is multiplied by e_i, and the products summed together to form the final output. Because the input to the system is composed of additive signal and noise terms, the sampled values X_i consist of additive signal and noise components. We can denote the noise components as the W_i, and the signal components may be denoted as θs_i. From (2-3) we see that the s_i here are the outputs of the single-pulse matched filter at times $t = i\Delta$, when the input is the pulse train $s(t)$ of (2-2).

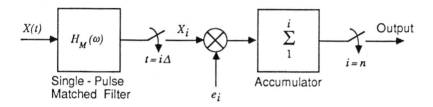

Figure 2.2 Linear System Maximizing Output SNR in Pulse Train Detection

If the noise process $W(t)$ in (2-3) is Gaussian, the system of Figure 2.2 leads to the *optimum* scheme for signal detection, that is, for testing the null hypothesis that $\theta = 0$ versus the alternative that $\theta > 0$. This is one of the central results of classical signal detection theory for Gaussian noise. The optimum detector uses the output of the linear system of Figure 2.2 as the test statistic, and compares it to a threshold to make a decision regarding the presence or absence of the signal.

The optimum detection scheme of Figure 2.2 can be given a particularly useful interpretation under some assumptions on the nature of the pulse train $s(t)$ and on the noise spectral density function $\Phi_W(\omega)$. Consider first the case where the noise is white, with a flat spectral density $\Phi_W(\omega) = N_0/2$. Then the impulse response $h_M(t)$ of the single-pulse matched filter is simply

$$h_M(t) = \frac{2}{N_0} p(\Delta - t) \qquad (2\text{-}6)$$

which is zero outside the interval $[0, \Delta]$, since $p(t)$ is confined entirely to this interval for non-overlapping pulses in $s(t)$. The covariance function $\rho_M(t)$ of the noise process at the output of the matched filter is then

$$\rho_M(t) = \frac{N_0}{2} h_M(t) * h_M(-t) \qquad (2\text{-}7)$$

which is therefore zero for $|t| \geq \Delta$. Thus the noise components W_i are *uncorrelated*, and hence independent for Gaussian noise.

More generally, the spectral density $\Phi_M(\omega)$ of the noise at the output of the single-pulse matched filter is

$$\Phi_M(\omega) = \Phi_W(\omega) \mid H_M(\omega) \mid^2$$
$$= P(\omega) \frac{P^*(\omega)}{\Phi_W(\omega)} \qquad (2\text{-}8)$$

so that

$$\rho_M(t) = p(t) * h_M(t + \Delta) \qquad (2\text{-}9)$$

Now the basic pulse $p(t)$ may in general occupy only a portion of the period Δ of the pulse train; let us therefore consider the case where $p(t)$ is non-zero only on the interval $[0, d]$, with $d \leq \Delta$. If the noise is not white the impulse response $h_M(t)$ of the single-pulse matched filter may have an effective duration which is larger than d. Let us now impose the condition, which will usually be satisfied for narrow pulses at least, that $h_M(t)$ is essentially zero outside of an interval of length $\Delta + (\Delta - d) = \Delta + \beta$. An examination of the functions shown in Figure 2.3 reveals that $\rho_M(t)$ will then be essentially zero for $|t| \geq \Delta$, so that we may again assume that the noise components W_i are uncorrelated and therefore independent for Gaussian noise. Under the same assumption the signal components s_i at the matched filter outputs will be simply

$$s_i = e_i \int_0^d p(t)\, h_M(\Delta - t)\, dt$$
$$= e_i \frac{1}{2\pi} \int_{-\infty}^{\infty} \frac{|P(\omega)|^2}{\Phi_W(\omega)}\, d\omega \qquad (2\text{-}10)$$

We may assume without any loss of generality that the pulse $p(t)$ has been defined with a scale factor making the quantity multiplying e_i in (2-10) equal to *unity*. We can then describe the samples X_i as

$$X_i = \theta e_i + W_i \qquad (2\text{-}11)$$

where the W_i are *independent and identically distributed* (i.i.d.) zero-mean Gaussian random variables.

The system of Figure 2.2 can be interpreted as being composed of two parts, a *single-pulse matched filter* generating the X_i and a *linear correlator* detector forming the test statistic by correlating the X_i with the e_i. Of course it is well-known that for X_i

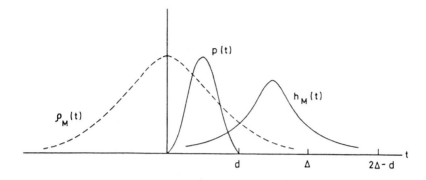

Figure 2.3 Filter Impulse Response and Output
Autocovariance Function, of Matched Filter
for Pulse Train Detection

given by (2-11) with i.i.d. Gaussian noise the linear-correlator detector is the uniformly most powerful (UMP) detector for $\theta > 0$.

What we have seen so far is that the linear processor maximizing the output SNR in detection of a pulse train may be considered quite generally to be operating on a set of intermediate observations X_i which are modeled by (2-1). Under a specific condition on the pulse and spectral density characteristics, the W_i in (2-1) are uncorrelated noise terms. If the noise process $W(t)$ is not Gaussian, the maximum output SNR scheme will not lead to an optimum detector. However, explicit solutions for the optimum detector are then also hard to come by. The use of a matched filter for SNR optimization is a common, simple, and generally well-founded engineering technique, and it usually makes good sense to continue to use it to generate the intermediate observations X_i even when the input noise is not Gaussian. However, as discussed at the end of Section 3.3 in Chapter 3, the use of linear predetection filtering is *not* appropriate when the noise process $W(t)$ contains impulsive non-Gaussian components. In the following development we will concentrate on the best use of the X_i modeled by (2-1) when the noise components W_i are not Gaussian.

2.2.2 Statistical Assumptions

Although the detection of a pulse train is only a particular example we chose to focus on, it illustrates the importance of the basic model of (2-1). We will now become more specific, and assume that the W_i form a sequence of i.i.d. random variables.

Their common univariate cumulative distribution function (cdf) will be denoted by F. Note that even though we do not require that F be Gaussian, we will assume that the W_i are independent. We will make some regularity assumptions about F that are generally met for all cases of interest. For future reference we will call these *Assumptions A*:

A. (i) F is absolutely continuous, so that a probability density function (pdf) f exists for the W_i;

(ii) f is absolutely continuous, so that its derivative $\dfrac{df(x)}{dx} = f'(x)$ exists for almost all x;

(iii) f' satisfies $\int_{-\infty}^{\infty} |f'(x)| \, dx < \infty$.

Assumptions A (i) and A (ii) are smoothness assumptions, and A (iii) is a technical requirement. These assumptions will allow us to justify certain mathematical operations, such as interchanges in the order of differentiation with respect to a parameter and integration of a function, although for the most part we will not provide details of such proofs.

This is also an appropriate place to introduce a second assumption, which we will call *Assumption B*:

B. $$I(f) \triangleq \int_{-\infty}^{\infty} \left[\frac{f'(x)}{f(x)} \right]^2 f(x) \, dx$$

$$< \infty \tag{2-12}$$

The quantity $I(f)$ is called *Fisher's information* for location shift, and the above assumption says that f has *finite* Fisher's information for location shift. The significance of $I(f)$ will become clear later in this chapter. We will find that this assumption is also satisfied by most noise density functions of interest. Notice that Assumption B implies that $f'(X_i)/f(X_i)$ has a finite variance under the noise-only condition $\theta = 0$; its mean value then is zero.

In this book we are concerned with problems where the noise density function f is completely specified, as a special case of the general *parametric* problem where f may have a finite number of unknown parameters (such as the noise variance). Our detection problem can be formulated as a statistical hypothesis-testing problem of choosing between a *null hypothesis* H_1 and an *alternative hypothesis* K_1 describing the joint density function f_X of the observation vector **X**, with

$$H_1: f_\mathbf{X}(\mathbf{x}) = \prod_{i=1}^{n} f(x_i) \tag{2-13}$$

$$K_1: f_\mathbf{X}(\mathbf{x}) = \prod_{i=1}^{n} f(x_i - \theta s_i), \text{ s specified, any } \theta > 0 \tag{2-14}$$

Here **s** is the vector $(s_1, s_2, \ldots, s_n)$ of signal components. Note that we are considering parametric hypotheses which completely define $f_\mathbf{X}$ to within a finite number of unknown parameters (here with only $\theta > 0$ unknown under the alternative hypothesis). Let us now proceed to obtain the structures of tests for H_1 versus K_1.

2.3 Locally Optimum Detection and Asymptotic Optimality

Since the alternative hypothesis K_1 is not a simple hypothesis, the signal amplitude value being unspecified, we cannot apply directly the fundamental lemma of Neyman and Pearson to obtain the structure of the optimum detector for our detection problem. For non-Gaussian noise densities it is also generally impossible to obtain UMP tests for the composite alternative hypothesis K_1.

To illustrate the difficulty, consider the special case where f is specified to be the *double-exponential* noise density function defined by

$$f(x) = \frac{a}{2} e^{-a|x|}, \quad a > 0 \tag{2-15}$$

The likelihood ratio for testing H_1 versus K_1, for a particular value $\theta = \theta_0 > 0$, is

$$L(\mathbf{X}) = \prod_{i=1}^{n} \frac{f(X_i - \theta_0 s_i)}{f(X_i)} \tag{2-16}$$

This now becomes

$$L(\mathbf{X}) = e^{-a \sum_{i=1}^{n} (|X_i - \theta_0 s_i| - |X_i|)} \tag{2-17}$$

giving

$$\ln L(\mathbf{X}) = a \sum_{i=1}^{n} (|X_i| - |X_i - \theta_0 s_i|) \tag{2-18}$$

Thus for *given* $\theta = \theta_0$, the test based on

$$\tilde{\lambda}(\mathbf{X}) = \sum_{i=1}^{n} (\,|X_i| - |X_i - \theta_0 s_i|\,) \qquad (2\text{-}19)$$

is an optimum test, since the constant a is positive. The optimum detector therefore has a test function defined by

$$\delta(\mathbf{X}) = \begin{cases} 1, & \tilde{\lambda}(\mathbf{X}) > t \\ r, & \tilde{\lambda}(\mathbf{X}) = t \\ 0, & \tilde{\lambda}(\mathbf{X}) < t \end{cases} \qquad (2\text{-}20)$$

where the threshold t and randomization probability r are chosen to obtain the desired value for the false-alarm probability p_f, so that the equation

$$E\{\delta(\mathbf{X}) \mid H_1\} = p_f \qquad (2\text{-}21)$$

is satisfied. Notice that we do not need randomization at $\tilde{\lambda}(\mathbf{X}) = t$ if this event has zero probability under H_1.

We can express $\tilde{\lambda}(\mathbf{X})$ of (2-19) in the form

$$\tilde{\lambda}(\mathbf{X}) = \sum_{i=1}^{n} l(X_i; \theta_0 s_i) \qquad (2\text{-}22)$$

where the characteristic l is defined by

$$l(x; \theta s) = |x| - |x - \theta s| \qquad (2\text{-}23)$$

This is shown in Figure 2.4 as a function of x and depends strongly on θ, so that $\tilde{\lambda}(\mathbf{X})$ cannot be expressed in a simpler form decoupling θ_0 and the X_i. For an implementation of the test statistic $\tilde{\lambda}(\mathbf{X})$ the value θ_0 of θ must be known, and a UMP test does not exist for this problem for $n > 1$.

One approach we might take in the above case is to use a *generalized likelihood ratio* (GLR) test, here obtained by using as the test statistic $\tilde{\lambda}(\mathbf{X})$ of (2-19) with θ_0 replaced by its *maximum likelihood* (ML) estimate under the alternative hypothesis K_1. This maximum-likelihood estimate $\hat{\theta}_{ML}$ is given implicitly as the solution of the equation

$$\sum_{i=1}^{n} s_i \, \mathrm{sgn}(X_i - \hat{\theta}_{ML} s_i) = 0 \qquad (2\text{-}24)$$

where

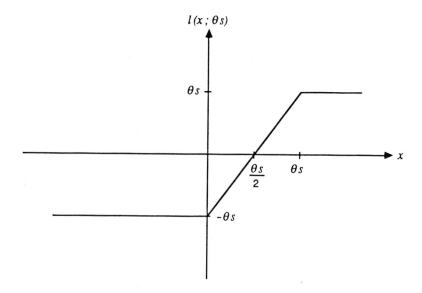

Figure 2.4 The Characteristic $l(x\,;\,\theta s)$ of Equation (2-23)

$$\text{sgn}(x) = \begin{cases} 1, & x > 0 \\ 0, & x = 0 \\ -1, & x < 0 \end{cases} \qquad (2\text{-}25)$$

provided that the solution turns out to be non-negative; otherwise, $\hat{\theta}_{ML} = 0$ (Problem 2.1). Thus the implementation of the GLR test is not simple. In addition, the distribution of the GLR test statistic under the null hypothesis is not easily obtained.

In the general case, for any noise density function f, the optimum detector for *given* $\theta = \theta_0 > 0$ under K_1 can be based on the test statistic

$$\begin{aligned}\lambda(\mathbf{X}) &= \ln L\,(\mathbf{X}) \\ &= \sum_{i=1}^{n} \ln \frac{f\,(X_i - \theta_0 s_i)}{f\,(X_i)}\end{aligned} \qquad (2\text{-}26)$$

which is of the form of $\tilde{\lambda}(\mathbf{X})$ of (2-22). But again, θ_0 must be specified and the detector will be optimum only for a signal with that amplitude. The GLR detector can be obtained if the ML estimate $\hat{\theta}_{ML}$ of θ can be found under the constraint that $\hat{\theta}_{ML}$ be non-negative. Once again, in general this will not lead to an easily implemented and easily analyzed system.

2.3.1 Locally Optimum Detectors

The above discussion shows that we have to search further in order to obtain reasonable schemes for detection of a known signal of unspecified amplitude in additive non-Gaussian noise. By a "reasonable" scheme we mean a detector that is practical to implement and relatively easy to analyze for performance, which should be acceptable for the anticipated range of input signal amplitudes. Fortunately, there is one performance criterion with respect to which it is possible to derive a simple and useful canonical structure for the optimum detector for our detection problem. This is the criterion of *local* detection power, and leads to detectors which are said to be *locally optimum*.

A *locally optimum* (LO) or locally most powerful detector for our problem is one which maximizes the *slope* of the detector power function at the *origin* ($\theta = 0$), from among the class of all detectors which have its false-alarm probability. Let Δ_α be the class of all detectors of size α for H_1 versus K_1. In our notation any detector D in Δ_α is based on a test function $\delta(\mathbf{X})$ for which

$$E\{\delta(\mathbf{X}) \mid H_1\} = \alpha \qquad (2\text{-}27)$$

Let $p_d(\theta \mid D)$ be the power function of detector D, that is,

$$p_d(\theta \mid D) = E\{\delta(\mathbf{X}) \mid K_1\} \qquad (2\text{-}28)$$

Formally, an LO detector D_{LO} of size α is a detector in Δ_α which satisfies

$$\max_{D \in \Delta_\alpha} \frac{d}{d\theta} p_d(\theta \mid D) \bigg|_{\theta=0} = \frac{d}{d\theta} p_d(\theta \mid D_{LO}) \bigg|_{\theta=0} \qquad (2\text{-}29)$$

It would be appropriate to use a locally optimum detector when one is interested primarily in detecting *weak* signals, for which θ under the alternative hypothesis K_1 remains close to zero. The idea is that an LO detector has a larger slope for its power function at $\theta = 0$ than any other detector D of the same size which is not an LO detector, and this will ensure that the power of the LO detector will be larger than that of the other detector at least for θ in some non-null interval $(0, \theta_{\max})$, with $\theta_{\max}$ depending on D. This is illustrated in Figure 2.5. Note that if an LO detector is not unique, then one may be better than another for $\theta > 0$. There is good reason to be concerned primarily with weak-signal detection. It is the weak signal that one has the most difficulty in detecting, whereas most *ad hoc* detection schemes should perform adequately for strong signals; after all, the detection probability is upper bounded by unity.

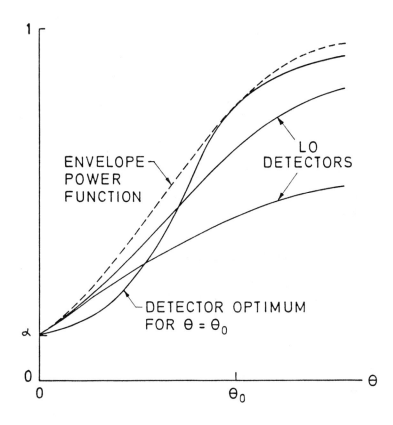

Figure 2.5 Power Functions of Optimum and LO Detectors

To obtain explicitly the canonical from of the LO detector for our problem, we can apply the generalized Neyman-Pearson lemma of Section 1.4, Chapter 1. Now the power function of a detector D based on a test function $\delta(\mathbf{X})$ is

$$p_d(\theta \mid D) = \int_{\mathbb{R}^n} \delta(\mathbf{x}) \prod_{i=1}^{n} f(x_i - \theta s_i) \, d\mathbf{x} \qquad (2\text{-}30)$$

where the integration is over the n-dimensional Euclidean space $\mathbb{R}^n$. The regularity Assumptions A allow us to get

$$\frac{d}{d\theta} p_d(\theta \mid D)\bigg|_{\theta=0} = \int_{R^n} \delta(\mathbf{x}) \frac{d}{d\theta} \prod_{i=1}^{n} f(x_i - \theta s_i)\bigg|_{\theta=0} d\mathbf{x}$$

$$= \int_{R^n} \delta(\mathbf{x}) \left[\sum_{i=1}^{n} - s_i \frac{f'(x_i)}{f(x_i)}\right] \prod_{i=1}^{n} f(x_i) \, d\mathbf{x}$$

$$= E\left\{\delta(\mathbf{X}) \left[\sum_{i=1}^{n} - s_i \frac{f'(X_i)}{f(X_i)}\right] \bigg| H_1\right\} \quad (2\text{-}31)$$

From this it follows, from the generalized Neyman-Pearson lemma, that a locally optimum detector D_{LO} is based on the test statistic

$$\lambda_{LO}(\mathbf{X}) = -\sum_{i=1}^{n} s_i \frac{f'(X_i)}{f(X_i)}$$

$$= \sum_{i=1}^{n} s_i \, g_{LO}(X_i) \quad (2\text{-}32)$$

where g_{LO} is the function defined by

$$g_{LO}(x) = \frac{-f'(x)}{f(x)} \quad (2\text{-}33)$$

Note that we may express $\lambda_{LO}(\mathbf{X})$ as

$$\lambda_{LO}(\mathbf{X}) = \sum_{i=1}^{n} \frac{d}{d\theta} \ln f(X_i - \theta s_i)\bigg|_{\theta=0}$$

$$= \frac{d}{d\theta} \sum_{i=1}^{n} \ln \frac{f(X_i - \theta s_i)}{f(X_i)}\bigg|_{\theta=0} \quad (2\text{-}34)$$

from which the LO detector test statistic (multiplied by θ) is seen to be a first-order approximation of the optimum detector test statistic given by (2-26).

The development of detection schemes for weak-signal situations can be traced back to the works of Middleton [1960] and Capon [1961]. Middleton used an expansion of the likelihood ratio in a Taylor series, from which an LO detection statistic is obtained as a weak-signal approximation. Capon considered explicitly the property of the LO detector as an optimum detector maximizing the slope of the power function for vanishing signal strength. An elaboration of Middleton's initial work is given in [Middleton, 1966]. Other early works employing the idea of a series expansion of the likelihood-ratio are those of Rudnick [1961],

Algazi and Lerner [1964], and Antonov [1967a, 1967b]. Helstrom [1967, 1968] also briefly discusses "threshold" or LO detection.

2.3.2 Generalized Correlator Detectors

The LO detector test statistic is, of course, not dependent on θ and has an easily implemented form. Let us define a *generalized correlator* (GC) test statistic $T_{GC}(\mathbf{X})$ as a test statistic of the form

$$T_{GC}(\mathbf{X}) = \sum_{i=1}^{n} a_i \, g(X_i) \tag{2-35}$$

where the a_i, $i = 1,2,...,n$ are a set of constants which are *correlated* with the $g(X_i)$, $i = 1,2,...,n$ to form $T_{GC}(\mathbf{X})$. The characteristic g is a memoryless or instantaneous nonlinearity. Then it is clear that $\lambda_{LO}(\mathbf{X})$ is a GC test statistic for which g is the locally optimum nonlinearity g_{LO} and the coefficients a_i are the known signal components s_i. Figure 2.6 shows the structure of a GC detector, from which it is clear that its implementation is quite easy. To set the threshold for any desired false-alarm probability p_f, the distribution of $T_{GC}(\mathbf{X})$ under the null hypothesis is required. The fact that $T_{GC}(\mathbf{X})$ is a linear combination of the i.i.d. $g(X_i)$ under the null hypothesis leads to some simplification of the problem of threshold computation. In practice the null-hypothesis distribution of $T_{GC}(\mathbf{X})$ may be computed through numerical convolution of the density functions of the $s_i g(X_i)$. If n is large enough, the Gaussian approximation may be used for the distribution of $T_{GC}(\mathbf{X})$, as we will see later.

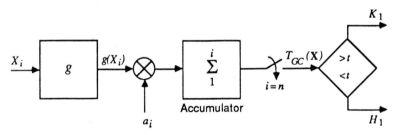

Figure 2.6 Generalized Correlator (GC) Detector

For f a zero-mean *Gaussian* density with variance σ^2, we have

$$g_{LO}(x) = \frac{x}{\sigma^2} \tag{2-36}$$

so that the locally optimum detector is based on a *linear correlator* (LC) test statistic

$$T_{LC}(\mathbf{X}) = \sum_{i=1}^{n} s_i X_i \qquad (2\text{-}37)$$

This is obtained from $\lambda_{LO}(\mathbf{X})$ of (2-32) using (2-36), after dropping the constant σ^2. This linear correlator detector based on $T_{LC}(\mathbf{X})$ is known to be not just locally optimum but also UMP for $\theta > 0$ when the noise is Gaussian. Clearly the LC test statistic is a special case of a GC test statistic.

For the double-exponential noise density of (2-15) we find that g_{LO} is given by

$$g_{LO}(x) = a\ \text{sgn}(x) \qquad (2\text{-}38)$$

Note that the *optimum* detector for $\theta = \theta_0$ in this case is based on the test statistic $\tilde{\lambda}(\mathbf{X})$ of (2-19). An equivalent detector is one based on $a\tilde{\lambda}(\mathbf{X})/\theta_0$ (since θ_0 and a are assumed known), which converges to $\lambda_{LO}(\mathbf{X})$ of (2-32) as $\theta_0 \to 0$. The LO detector for the double-exponential density is therefore a *sign correlator* (SC) detector based on the GC statistic

$$T_{SC}(\mathbf{X}) = \sum_{i=1}^{n} s_i\ \text{sgn}(x_i) \qquad (2\text{-}39)$$

In the case of constant signals we may set the s_i to be unity; the result is what is called the *sign detector*, of which the SC detector is a general form. (We will study the sign and SC detectors in detail as *nonparametric* detectors in a sequel to this book.) Their property of being *robust* detectors is also well known.

The LC and the SC detectors (and their special cases for $s_i = 1$, all i, the *linear* and *sign* detectors, respectively) are the best-known of the LO detectors for known signals in additive i.i.d. noise. We will, in fact, encounter other versions of these detectors in Chapters 5, 6, and 7. The LC detector is optimum for the important special case of Gaussian noise, and therefore is used as a standard against which other detectors are compared for performance under Gaussian noise. The SC detector is a particularly simple example of a detector which is optimum for a non-Gaussian noise density, which does happen to be a useful simple model of non-Gaussian noise in several applications. Its very simple structure together with its nonparametric and robust performance characteristics make it, too, a commonly used standard against which other detectors are compared for performance and complexity.

2.3.3 Asymptotic Optimality

For our detection problem formalized as that of testing H_1 versus K_1, a locally optimum detector D_{LO} of size α has a power function $p_d(\theta \mid D_{LO})$ for which the slope $p_d{}'(\theta \mid D_{LO})$ at $\theta = 0$ is a maximum from among all detectors D of size α. In the non-Gaussian situation an LO detector will generally not be a UMP detector. For some *particular* value $\theta_0 > 0$ for θ we can find the *optimum* (Neyman-Pearson) detector D_{NP} of size α; let its power function be explicitly written as $p_d(\theta \mid \theta_0, D_{NP})$. Then in general we will have

$$p_d(\theta_0 \mid D_{LO}) < p_d(\theta_0 \mid \theta_0, D_{NP}) \qquad (2\text{-}40)$$

for $0 < \alpha < 1$, and this will be true for θ_0 arbitrarily small.

Let us call the function of θ_0 on the right-hand side of (2-40) the *envelope power function*, which is therefore the function $p_d(\theta \mid E)$ defined by

$$p_d(\theta \mid E) = p_d(\theta \mid \theta, D_{NP}) \qquad (2\text{-}41)$$

Clearly, we will find that $\lim_{\theta \to 0} p_d(\theta \mid E) = \alpha$. Since we have assumed the non-existence of a UMP detector, no *single* detector exists which has the power function $p_d(\theta \mid E)$. Rather, the envelope power function now becomes the standard against which the performances of other detectors may be compared. The relationship between these power functions is illustrated in Figure 2.5.

In using an LO detector we can be assured only that $p_d{}'(0 \mid D)$ is maximized. In justifying the use of an LO detector we have mentioned that the case of weak signals implied by the condition $\theta \to 0$ is an important case in practical applications. On the other hand note that for any *fixed* sample size n, the power or detection probability approaches the value α (the false-alarm probability) for $\theta \to 0$. Thus while the condition $\theta \to 0$ is of considerable interest, it is generally used *in conjunction with* an assumption that $n \to \infty$. The practical significance is that in detection of a weak signal one necessarily uses a relatively large sample size to get a reasonable value for the probability of correct detection. We see that with local optimality we addressed only one part (the "local signal" case) of this combined asymptotic situation. We will now consider the combined asymptotic assumption $\theta \to 0$ and $n \to \infty$, and show that in general the LO detector has an optimality property for this asymptotic case.

The problem can be formally stated as a *sequence* of hypothesis-testing problems, $\{H_{1,n}$ versus $K_{1,n}, \; n = 1,2,...\}$ where $H_{1,n}$ is simply an explication of the fact that H_1 of (2-13) is applicable for the case of n independent observations from the noise density. The sequence of alternatives $\{K_{1,n}, \; n = 1,2,...\}$ is described by

$$K_{1,n}: \; f_X(\mathbf{x}) = \prod_{i=1}^{n} f(x_i - \theta_n s_i), \quad n = 1,2,...,$$

$$\theta_n > 0, \theta_n \to 0 \quad . \quad (2\text{-}42)$$

We shall be interested in particular types of sequences $\{\theta_n\}$ converging to zero. For example, we could have $\theta_n = \gamma/\sqrt{n}$ for some fixed $\gamma > 0$, so that $n\theta_n^2$ remains fixed at γ^2 as $n \to \infty$. In the *constant signal* case ($s_i = 1$, all n) this models a sequence of detection problems with increasing sample sizes and decreasing signal amplitudes in such a way that the *total signal energy* $\theta_n^2 \sum_{i=1}^{n} s_i^2 = n\theta_n^2$ remains fixed. In the *time-varying signal* case, the condition that $\frac{1}{n} \sum_{i=1}^{n} s_i^2$ converges (without loss of generality, to unity) as $n \to \infty$ allows a similar interpretation to be made in the asymptotic case. Let us, in fact, henceforth make such an assumption about our known signal sequence $\{s_i, \; i = 1,2,...\}$, and also require it to be uniformly bounded for mathematical convenience. These may be stated as *Assumptions C*:

C. (i) There exists a finite, non-zero bound U_s such that

$$0 \leq |s_i| \leq U_s, \quad i = 1,2,... \quad (2\text{-}43)$$

(ii) The asymptotic average signal power is finite and non-zero,

$$0 < \lim_{n \to \infty} \frac{1}{n} \sum_{i=1}^{n} s_i^2 = P_s^2 < \infty \quad (2\text{-}44)$$

We should note one more technical detail. In defining the $K_{1,n}$ by (2-42), there is the implication that as n increases the known-signal vector $\mathbf{s} = (s_1, s_2, ..., s_n)$ grows by having *additional* components appended to it, without change in the *existing* components. This appropriately models a situation where the observations X_i are obtained as samples taken at a *fixed* sampling rate from some continuous-time waveform, with the observation time *increasing* to generate the sequence of hypothesis-testing problems. On the other hand, one may have a situation where the observation interval is *fixed*, with the sampling rate *increasing* to generate the sequence of hypothesis-testing problems. Such cases are handled by replacing the s_i in (2-42) with $s_{i,n}$, thus explicitly allowing the set of n signal components under $K_{1,n}$ to depend more generally on n.

The idea in considering formally a *sequence* of hypothesis-testing problems $\{H_{1,n} \text{ versus } K_{1,n}, n = 1,2,...\}$ is the following: if we can find a corresponding *sequence* $\{D_n, n = 1,2,...\}$ of detectors which in the limit $n \to \infty$ have some optimality property, then for sufficiently large finite n the use of D_n will give approximately optimum performance.

Since we are considering a sequence of hypothesis-testing problems indexed by n, we will now modify our notation slightly and use, for instance, $p_d(\theta_n \mid D_n)$ for the power function of a size-α detector D_n based on n observations $X_1, X_2, ..., X_n$ for $H_{1,n}$ versus $K_{1,n}$. What we would like to do is consider sequences $\{\theta_n\}_{n=1}^\infty$ converging to zero is such a way that the envelope power function values $p_d(\theta_n \mid E_n)$ converge to values which lie strictly between α and 1 (for size-α detectors). Then we would be looking at a sequence of detection problems in which the sample size is growing and signal amplitude is shrinking in such a way as to allow the optimum detectors always to yield a useful level of asymptotic detection performance (detection probability larger than α) but without allowing asymptotically "perfect" detection (detection probability less than unity).

Definition: We will say that a sequence of detectors $\{D_{AO,n}, n = 1,2,...\}$ is *asymptotically optimum* (AO) at level α for $\{H_{1,n} \text{ versus } K_{1,n}, n = 1,2,...\}$ if [1]

(i) $$\lim_{n \to \infty} E\{\delta_{AO,n}(\mathbf{X}) \mid H_{1,n}\} \qquad (2\text{-}45)$$
$$= \lim_{n \to \infty} p_d(0 \mid D_{AO,n}) \leq \alpha$$

and

(ii) $$\lim_{n \to \infty} [p_d(\theta_n \mid E_n) - p_d(\theta_n \mid D_{AO,n})] = 0 \qquad (2\text{-}46)$$

Here $\delta_{AO,n}(\mathbf{X})$ is the test function for the detector $D_{AO,n}$.

Condition (i) above will obviously be satisfied if each detector $D_{A0,n}$ in the sequence is a size-α detector. According to condition (ii), an AO sequence of detectors has a sequence of power values for the alternatives $K_{1,n}$ defined by the θ_n which is in the *limit* $n \to \infty$ equal to the power of the optimum detector. If

---
[1] (i) May be made more general by replacing "$\lim_{n \to \infty}$" by "$\limsup_{n \to \infty}$"; this would not necessarily require $p_d(0 \mid D_{AO,n})$ to converge as $n \to \infty$.

θ_n, $n = 1,2,...$, defines a sequence of alternatives for which the sequence of optimum size-α detectors is *consistent*, then $\lim_{n \to \infty} p_d(\theta_n \mid E_n) = 1$. In this case *any* other sequence of detectors which is simply consistent (and has asymptotic size α) will be an AO sequence. Thus it will be of most interest to consider cases for which $\alpha < \lim_{n \to \infty} p_d(\theta_n \mid E_n) < 1$.

Note carefully that the property of being AO belongs to a *sequence* of detectors. Usually, however, we consider sequences of detectors employing exactly the same *rule* for computing the test statistic for each n; for example, we might consider a sequence of SC detectors with test statistics defined by (2-39) for each n. In such cases we often say that a particular detector is AO, when we really mean that the sequence of such detectors is AO. The property of being locally optimum, on the other hand, belongs to an *individual* detector (operating on some sample of fixed size n).

2.3.4 Asymptotic Optimality of LO Detectors

Let us consider the sequence $\{D_{LO,n}, n = 1,2,...\}$ of LO detectors of size α for our sequence of hypothesis-testing problems. Then we can show that this is an AO sequence of detectors for the alternatives $K_{1,n}$ defined by $\theta_n = \gamma/\sqrt{n}$, $n = 1,2,...$, for any fixed $\gamma > 0$. This optimality property of the sequence of LO detectors gives them a stronger justification for use when sample sizes are large. To prove the asymptotic optimality of any sequence of detectors $\{D_n\}$ we have to be able to obtain the limiting values of the powers $p_d(\theta_n \mid D_n)$ and $p_d(\theta_n \mid E_n)$, and these can be obtained if the asymptotic distributions of the test statistics are known. For the envelope power function $p_d(\theta_d \mid E_n)$ we have to consider the optimum detectors based on the log-likelihood functions $\lambda(\mathbf{X})$ of (2-26), and obtain their asymptotic distributions for $\theta_0 = \theta_n$ and $n \to \infty$. Let us now carry out this asymptotic analysis in a heuristic way, with any regularity conditions required to make our analysis rigorously valid implicitly assumed to hold.

The log-likelihood function $\lambda(\mathbf{X})$ of (2-26) with $\theta_0 = \gamma/\sqrt{n}$ may be expanded in the Taylor series

$$\lambda(\mathbf{X}) = \sum_{i=1}^{n} \ln \frac{f(X_i)}{f(X_i)} + \frac{\gamma}{\sqrt{n}} \sum_{i=1}^{n} s_i \left[\frac{-f'(X_i)}{f(X_i)} \right]$$

$$+ \frac{1}{2} \frac{\gamma^2}{n} \sum_{i=1}^{n} s_i^2 \left\{ \frac{f''(X_i)}{f(X_i)} - \left[\frac{f'(X_i)}{f(X_i)} \right]^2 \right\}$$

$$+ o(1)$$

$$\approx \frac{\gamma}{\sqrt{n}} \lambda_{LO}(\mathbf{X})$$

$$+ \frac{1}{2} \frac{\gamma^2}{n} \sum_{i=1}^{n} s_i^2 \left\{ \frac{f''(X_i)}{f(X_i)} - \left[\frac{f'(X_i)}{f(X_i)} \right]^2 \right\} \quad (2\text{-}47)$$

for large n. Note that a quantity Z_n is said to be of order $o(1/a_n)$ if $a_n Z_n \to 0$ (in probability). Thus the only essential way in which the above approximation differs from the LO test statistic $\lambda_{LO}(\mathbf{X})$ of (2-32) is in the second term above. But this term converges in probability,

$$\lim_{n \to \infty} \frac{\gamma^2}{2n} \sum_{i=1}^{n} s_i^2 \left\{ \frac{f''(X_i)}{f(X_i)} - \left[\frac{f'(X_i)}{f(X_i)} \right]^2 \right\}$$

$$= \frac{1}{2} \gamma^2 P_s^2 E \left\{ \frac{f''(X_i)}{f(X_i)} - \left[\frac{f'(X_i)}{f(X_i)} \right]^2 \right\}$$

$$= -\frac{1}{2} \gamma^2 P_s^2 I(f) \quad (2\text{-}48)$$

under both $H_{1,n}$ and $K_{1,n}$. We have assumed that $\int_{-\infty}^{\infty} f''(x) \, dx$ is zero. Note that P_s^2 is the average known-signal power and $I(f)$ is Fisher's information for location shift of (2-12). The first term $(\gamma/\sqrt{n})\lambda_{LO}(\mathbf{X})$ is asymptotically Gaussian under $H_{1,n}$, with mean value 0 and variance $\gamma^2 P_s^2 I(f)$. To find the asymptotic mean and variance of $(\gamma/\sqrt{n})\lambda_{LO}(\mathbf{X})$ under $K_{1,n}$, we expand it as

$$\frac{\gamma}{\sqrt{n}} \lambda_{LO}(\mathbf{X}) = \frac{\gamma}{\sqrt{n}} \sum_{i=1}^{n} s_i \left[\frac{-f'(W_i + \gamma/\sqrt{n} \ s_i)}{f(W_i + \gamma/\sqrt{n} \ s_i)} \right]$$

$$= \frac{\gamma}{\sqrt{n}} \sum_{i=1}^{n} s_i \left[\frac{-f'(W_i)}{f(W_i)} \right]$$

$$+ \frac{\gamma^2}{n} \sum_{i=1}^{n} s_i^2 \left\{ \frac{-f''(W_i)}{f(W_i)} + \left[\frac{f'(W_i)}{f(W_i)} \right]^2 \right\}$$

$$+ o(1) \quad (2\text{-}49)$$

We conclude that the asymptotic mean of $(\gamma/\sqrt{n})\lambda_{LO}(\mathbf{X})$ is $\gamma^2 P_s^2 I(f)$ and its asymptotic variance is also $\gamma^2 P_s^2 I(f)$ under $K_{1,n}$. Furthermore, $(\gamma/\sqrt{n})\lambda_{LO}(\mathbf{X})$ is also asymptotically normally distributed under the sequence of alternatives $K_{1,n}$.

To summarize, then, we have that the sequence of optimum detection test statistics converges in distribution to a test statistic with a Gaussian distribution with mean $-\frac{1}{2} \gamma^2 P_s^2 I(f)$ and variance $\gamma^2 P_e^2 I(f)$ under the null hypotheses, and to a test statistic with a Gaussian distribution with mean $\frac{1}{2}\gamma^2 P_e^2 I(f)$ and variance $\gamma^2 P_e^2 I(f)$ under the alternative hypotheses. The sequence of LO test statistics $(\gamma/\sqrt{n})\lambda_{LO}(\mathbf{x})$ is similarly asymptotically Gaussian, with mean zero and variance $\gamma^2 P_e^2 I(f)$ under the null hypotheses and with mean and variance both equal to $\gamma^2 P_e^2 I(f)$ under the alternative hypotheses. From this we conclude that for the optimum detectors we get, with Φ the standard Gaussian distribution function,

$$\lim_{n \to \infty} p_d\left(\frac{\gamma}{\sqrt{n}} \mid E_n\right) = \lim_{n \to \infty} P\{\lambda(\mathbf{X}) > t_n \mid K_{1,n}\}$$

$$= 1 - \Phi\left[\frac{t_\alpha - \frac{1}{2}\gamma^2 P_e^2 I(f)}{\gamma P_e \sqrt{I(f)}}\right] \quad (2\text{-}50)$$

where the thresholds t_n for size-α optimum detectors converge to t_α given by

$$\alpha = 1 - \Phi\left[\frac{t_\alpha + \frac{1}{2}\gamma^2 P_e^2 I(f)}{\gamma P_e \sqrt{I(f)}}\right] \quad (2\text{-}51)$$

This gives

$$\lim_{n \to \infty} p_d\left(\frac{\gamma}{\sqrt{n}} \mid E_n\right) = 1 - \Phi[\Phi^{-1}(1-\alpha) - \gamma P_e \sqrt{I(f)}] \quad (2\text{-}52)$$

Similarly, the sequence of LO detectors $D_{LO,n}$ using $(\gamma/\sqrt{n})\lambda_{LO}(\mathbf{X})$ as test statistics and a fixed threshold $t_\alpha + \frac{1}{2}\gamma^2 P_e^2 I(f)$ have asymptotic size α and asymptotic power

$$\lim_{n \to \infty} p_d\left(\frac{\gamma}{\sqrt{n}} \mid D_{LO,n}\right)$$

$$= \lim_{n \to \infty} P\left\{\frac{\gamma}{\sqrt{n}}\lambda_{LO}(\mathbf{X}) > t_\alpha + \frac{1}{2}\gamma^2 P_e^2 I(f) \mid K_{1,n}\right\}$$

$$= 1 - \Phi\left[\frac{t_\alpha - \frac{1}{2}\gamma^2 P_e^2 I(f)}{\gamma P_e \sqrt{I(f)}}\right]$$

$$= \lim_{n \to \infty} p_d\left(\frac{\gamma}{\sqrt{n}} \mid E_n\right) \quad (2\text{-}53)$$

This makes the sequence $\{D_{LO,n},\ n = 1,2,...\}$ an AO sequence of detectors for $\theta_n = \gamma/\sqrt{n}$ under $K_{1,n}$.

The above development does not constitute a rigorous proof of the fact that the sequence of LO detectors are AO for the alternatives $\theta_n = \gamma/\sqrt{n}$. A rigorous proof along the above lines would require additional specific regularity conditions to be imposed on the noise density functions f. Actually, it is possible to establish quite rigorously the required asymptotic normality results above *without* assuming regularity conditions beyond those of assumptions A, B, and C. It would not be appropriate to enter into the details of such a proof here, involving as it does some lengthy mathematical details. We shall observe here only that such a proof makes use of some very useful results known as LeCam's lemmas, which are given detailed exposure by Hajek and Sidak [1967, Ch. VI].

The sequence of LO detectors is not the only sequence of detectors which is AO for $\theta_n = \gamma/\sqrt{n}$. The sequence of *optimum* detectors is an obvious example of another AO sequence of detectors. Loosely speaking, any sequence of detectors with the correct asymptotic size and with test statistics which converge to LO test statistics as $n \to \infty$ will be an AO sequence of detectors. For example, consider the sequence of test statistics

$$T(\mathbf{X}) = \sum_{i=1}^{n} s_i\, q_n(X_i), \quad n = 1,2,\ldots \qquad (2\text{-}54)$$

where

$$q_n(x) = \begin{cases} 1, & x > \dfrac{1}{n} \\ 0, & -\dfrac{1}{n} \leq x \leq \dfrac{1}{n} \\ -1, & x < -\dfrac{1}{n} \end{cases} \qquad (2\text{-}55)$$

Then it can be shown easily that the sequence of size-α detectors based on this sequence of test statistics is AO for $\theta_n = \gamma/\sqrt{n}$ and for the double-exponential noise density. The LO detector for any specific value of the sample size may not be very convenient to implement. If we use instead a simpler detector for each n, defined in such a way that this sequence of detectors is asymptotically optimum, then for large n and small θ we will get performance very close to that of the LO detector.

We have given a careful and somewhat lengthy discussion of AO detectors in this section for several reasons. First, there has been some confusion in earlier work with regard to the asymptotic optimality properties of LO detectors. Second, we will find that this careful discussion will help us to better understand the *asymptotic relative efficiency* (ARE) as a relative performance measure for two (sequences of) detectors. Finally, having given this development of asymptotic optimality specifically for the known signal detection problem, we will feel justified in omitting its treatment for other types of detection problems. It is possible to obtain AO detectors for many other types of detection problems, including those formulated for continuous-time observations.

We have already referred to the book by Hajek and Sidak [1967] as an excellent, though somewhat advanced treatment of this topic. In signal detection applications the asymptotic optimality criterion has been discussed by Levin and Kushnir [1969], by Levin and Rybin [1969], by Levin, Kushnir, and Pinskiy [1971], and by Kutoyants [1975, 1976], among others. One topic we have omitted is that of AO detection in *correlated* noise, which would have taken us too far outside the intended scope of this book. The interested reader is referred to the works of Pinskiy [1971], Poor [1982], Poor and Thomas [1979, 1980], and Halverson and Wise [1980a, 1980b] as examples of such investigations.

2.4 Detector Performance Comparisons

We have noted that the linear correlator detector based on the test statistic $T_{LC}(\mathbf{X})$ of (2-37) is a UMP detector for the known-signal detection problem when the noise is Gaussian. When the noise is not Gaussian, however, the LO detector is a generalized correlator detector using test statistic $\lambda_{LO}(\mathbf{X})$ of (2-32), a special of $T_{GC}(\mathbf{X})$ of (2-35). Since the use of detectors which are optimum under Gaussian conditions is widespread, we will be particularly interested now in comparing the performance of the LC detector with other GC detectors for different noise density functions f, for testing H_1 versus K_1. For example, it would be useful to know how well the LC detector performs relative to the sign correlator (SC) detector when the noise has the double-exponential density function, and also how much better it is than the SC detector when the noise is, indeed, Gaussian.

In comparing the performances of two detectors D_A and D_B in testing H_1 versus K_1 the relevant quantities which have to be considered are the sample size n, the signal amplitude θ, the false-alarm probability p_f, the noise density function f, and the detection probabilities $p_d(\theta \mid D_A)$ and $p_d(\theta \mid D_B)$. For a given density function f, for example, one can look at different combinations of n and θ and obtain for each the relationship between p_f and the detection probability p_d, the *receiver operating characteristics* (ROC), for each detector. Alternatively, one may obtain for each detector the power functions $p_d(\theta \mid D)$ for fixed f and different combinations of the sample size n and false-alarm probability $p_f = \alpha$.

It can be appreciated that this can become a rather formidable task even for simple GC detectors. Consider, for example, the LC detector when the noise density function is not Gaussian. To obtain its power function for a given n and size α one has generally to resort to a numerical technique to find the distribution of the LC test statistic for the given n. The comparison is even more difficult for the GC test statistics using nonlinear functions g.

Even if it were easy to generate power functions or ROCs for the detectors being compared, there would be entire families of such performance functions, parameterized by n and α or n and θ, respectively, for each detector for any noise density function f. What would be very useful would be to have a real-valued relative performance index which can be derived as a function of f, summarizing some particularly relevant aspect of the exhaustive performance comparison. These considerations lead us to the introduction of the asymptotic relative efficiency (ARE) of two detectors as a measure of their relative performance.

2.4.1 Asymptotic Relative Efficiency and Efficacy

In Section 1.6 of Chapter 1 we gave the technical details about the ARE as a measure of the relative performance of two detectors. We now explain it and illustrate its use specifically for the known-signal detectors. The most important thing to note about the ARE is that it measures the relative performance of two detectors in the *asymptotic* case $n \to \infty$ for a *sequence* of hypothesis-testing problems. In the specific context of known-signal detection we consider the sequence $\{H_{1,n} \text{ versus } K_{1,n}, n = 1,2,...\}$. We have $\theta_n \to 0$ as $n \to \infty$, and θ_n is chosen in a manner which makes the test statistics of both detectors have well-defined asymptotic distributions (of the same type) as $n \to \infty$. If under $H_{1,n}$ the two sequences of test statistics also have well-defined asymptotic distributions of a common type, the *parameters* of these four asymptotic distributions may be used to characterize the asymptotic relative performance of the two detectors. Under the specific assumptions stated in Section 1.6 (asymptotic distributions are Gaussian, each detector has test statistics which have the same asymptotic variance under both $H_{1,n}$ and $K_{1,n}$), the asymptotic means and variances can be used to define the ARE as an index of relative performance. The ARE has a numerical value which can be obtained for given f.

Let us now get down to specifics for GC detectors. Consider in general the use of a GC detector using coefficients a_i, as described by the test statistic $T_{GC}(\mathbf{X})$ of (2-35). Since we will be considering a sequence of such GC detectors based on some fixed g, in the limiting case there will be an infinite number of coefficients a_i defining $T_{GC}(\mathbf{X})$. We make the following regularity *Assumptions D*:

D. (i) There exists a finite non-zero bound U_a such that

$$0 \leq |a_i| \leq U_a, \quad i = 1,2,... \tag{2-56}$$

(ii) The asymptotic average coefficient power is finite and non-zero,

$$0 < \lim_{n \to \infty} \frac{1}{n} \sum_{i=1}^{n} a_i^2 = P_a^2 < \infty \tag{2-57}$$

(iii) $g(X_i)$ has zero mean and finite variance under f,

$$\int_{-\infty}^{\infty} g(x) f(x) dx = 0 \tag{2-58}$$

and

$$\int_{-\infty}^{\infty} g^2(x) f(x) dx < \infty \tag{2-59}$$

Assumptions D (i) and (ii) are similar to Assumptions C (i) and (ii) of Section 2.3 for the known signal sequence, and are quite reasonable conditions to impose on the coefficients. Under D (iii) we have that $E\{g(X_i) | H_1\} = 0$, which is not restrictive since any arbitrary g can always be normalized to satisfy this condition, by subtracting from it its mean value under H_1. Finally, the finite variance assumption is also quite reasonable. Notice that the LO nonlinearity g_{LO} for given f does satisfy condition D (iii), for f satisfying Assumptions A and B of Section 2.2.

Now consider the sequence of hypothesis-testing problems $\{H_{1,n}$ versus $K_{1,n}, n = 1,2,...\}$, with $\theta_n = \gamma/\sqrt{n}$ for some fixed $\gamma > 0$, and the sequence of GC detectors $\{D_{GC,n}, n = 1,2,...\}$ based on test statistics $T_{GC}(\mathbf{X})$ given by (2-35) and satisfying assumptions D. Under the null hypotheses $H_{1,n}$ it is easy to see that $(\gamma/\sqrt{n}) T_{GC}(\mathbf{X})$ has mean zero and asymptotic variance

$$\lim_{n \to \infty} V\left\{ \frac{\gamma}{\sqrt{n}} T_{GC}(\mathbf{X}) | H_{1,n} \right\} = \gamma^2 P_a^2 \int_{-\infty}^{\infty} g^2(x) f(x) dx \tag{2-60}$$

Furthermore, it follows from a central limit theorem that $(\gamma/\sqrt{n}) T_{GC}(\mathbf{X})$ is asymptotically Gaussian under $H_{1,n}$.

To obtain the characteristics under $K_{1,n}$, let us use the expansion

$$\frac{\gamma}{\sqrt{n}} T_{GC}(\mathbf{X}) = \frac{\gamma}{\sqrt{n}} \sum_{i=1}^{n} a_i \, g\left(W_i + \frac{\gamma}{\sqrt{n}} s_i\right)$$

$$= \frac{\gamma}{\sqrt{n}} \sum_{i=1}^{n} a_i \, g(W_i) + \frac{\gamma^2}{n} \sum_{i=1}^{n} a_i s_i \, g'(W_i)$$

$$+ o(1) \qquad (2\text{-}61)$$

[cf. expansion (2-49)], assuming for the time being the additional regularity conditions required to make this valid. We find from this that

$$\lim_{n \to \infty} E\left\{\frac{\gamma}{\sqrt{n}} T_{GC}(\mathbf{X}) \mid K_{1,n}\right\} = \gamma^2 P_{as} \int_{-\infty}^{\infty} g'(x) f(x) \, dx$$

$$= -\gamma^2 P_{as} \int_{-\infty}^{\infty} g(x) f'(x) \, dx \qquad (2\text{-}62)$$

$$\lim_{n \to \infty} V\left\{\frac{\gamma}{\sqrt{n}} T_{GC}(\mathbf{X}) \mid K_{1,n}\right\} = \gamma^2 P_a^2 \int_{-\infty}^{\infty} g^2(x) f(x) \, dx$$

$$(2\text{-}63)$$

and that $(\gamma/\sqrt{n}) T_{GC}(\mathbf{X})$ is also asymptotically Gaussian under $K_{1,n}$. Here the quantity P_{as} is

$$P_{as} = \lim_{n \to \infty} \frac{1}{n} \sum_{i=1}^{n} a_i s_i \qquad (2\text{-}64)$$

This heuristic proof that $(\gamma/\sqrt{n}) T_{GC}(\mathbf{X})$ is asymptotically Gaussian with mean $-\gamma^2 P_{as} \int gf'$ and variance $\gamma^2 P_a \int g^2 f$ follows a line of reasoning similar to one that we employed for the LO detectors earlier. Note that for $g = g_{LO}$ and $a_i = s_i$ these results reduce to those obtained for the LO detectors. Furthermore, it is also true that the asymptotic normality under $K_{1,n}$ with the mean and variance we have given above can be established *without* using any regularity assumptions other than Assumptions A, B, C, and D. Such a proof depends on use of LeCam's lemmas, which, as we have indicated earlier, are discussed in Hajek and Sidak [1967, Ch. VI].

Proceeding as we have done before for the LO detectors, we find that for the sequence of GC detectors with asymptotic size α

the asymptotic power function is (Problem 2.3)

$$\lim_{n \to \infty} p_d\left(\frac{\gamma}{\sqrt{n}} \mid D_{GC,n}\right)$$
$$= 1 - \Phi\left[\Phi^{-1}(1-\alpha) - \gamma\frac{P_{as}}{P_a} E(g,f)\right] \quad (2\text{-}65)$$

where

$$E(g,f) = \frac{-\int_{-\infty}^{\infty} g(x)f'(x)\, dx}{\left[\int_{-\infty}^{\infty} g^2(x)f(x)\, dx\right]^{1/2}}$$

$$= \frac{\int_{-\infty}^{\infty} g(x)\left[\frac{-f'(x)}{f(x)}\right] f(x)\, dx}{\left[\int_{-\infty}^{\infty} g^2(x)f(x)\, dx\right]^{1/2}} \quad (2\text{-}66)$$

The function $1 - \Phi[\Phi^{-1}(1-\alpha) - c]$ is an increasing function of c with value α at $c = 0$. Thus for $\gamma > 0$ the limiting detection power will be no less than α if $P_{as} E(g,f)$ is non-negative, making the GC detectors asymptotically unbiased. This quite reasonable requirement will always be assumed to hold; the sign of g or of the coefficient sequence can always be picked to meet this requirement. Comparing the limiting power given by (2-52) for the AO detectors and the result (2-65) for the GC detectors, we see that an index of their asymptotic relative performance is provided by the following ratio, which we will call the *asymptotic relative gain* (ARG):

$$ARG_{GC,AO} = \frac{P_{as} E(g,f)}{P_a P_s \sqrt{I(f)}}$$
$$= \rho_{as} \frac{E(g,f)}{\sqrt{I(f)}} \quad (2\text{-}67)$$

Here ρ_{as} is the correlation coefficient

$$\rho_{as} = \frac{P_{as}}{P_a P_s} \quad (2\text{-}68)$$

The coefficient ρ_{as} attains its maximum value of unity if $a_i = ks_i$, all i, for any $k > 0$. We also find that

$$\frac{E(g,f)}{\sqrt{I(f)}} = \frac{\int_{-\infty}^{\infty} g(x)g_{LO}(x)f(x)\,dx}{\left[\int_{-\infty}^{\infty} g^2(x)f(x)\,dx \int_{-\infty}^{\infty} g_{LO}^2(x)f(x)\,dx\right]^{1/2}}$$

$$\leq 1 \qquad (2\text{-}69)$$

with equality being achieved when $g \equiv g_{LO}$.

Let us view the right-hand sides of (2-52) and (2-65) as approximations for the actual detection probabilities for the AO and GC detectors, respectively, for large sample sizes. Then we may write

$$p_d(\theta \mid D_{GC,n}) \approx 1 - \Phi[\Phi^{-1}(1-\alpha) - \theta\sqrt{n}\ P_s \rho_{as} E(g,f)] \qquad (2\text{-}70)$$

and

$$p_d(\theta \mid D_{AO,n}) \approx 1 - \Phi[\Phi^{-1}(1-\alpha) - \theta\sqrt{n}\ P_s \sqrt{I(f)}] \qquad (2\text{-}71)$$

For fixed n (and P_s, since the signal is known) we see that the $ARG_{GC,AO}$ measures the relative *signal amplitudes* required for the performances of the two detectors to be identical. Specifically, $ARG_{GC,AO}$ is the amplitude of the signal required by the AO detector relative to that required by the GC detector for their detection probabilities to be the same at any common value α of the false-alarm probability, in the asymptotic case of large n.

On the other hand, if we fix θ at some small value, then it follows immediately that the *square* of $ARG_{GC,AO}$ is the $ARE_{GC,AO}$, the ratio of *sample sizes* required by the two detectors for identical asymptotic performance. Thus we have

$$\begin{aligned}ARE_{GC,AO} &= [ARG_{GC,AO}]^2 \\ &= \frac{(P_{as}^2/P_a^2)\,E^2(g,f)}{P_s^2 I(f)}\end{aligned} \qquad (2\text{-}72)$$

The numerator in (2-72) may be identified as the *efficacy* ξ_{GC} of the GC detector,

$$\xi_{GC} = \rho_{as}^2 P_s^2 E^2(g,f) \qquad (2\text{-}73)$$

which reduces to $P_s^2 I(f)$ for the AO detector using $a_i = s_i$ and $g \equiv g_{LO}$. Since the signal sequence is known we will henceforth assume that $a_i = s_i$, so that we get $\rho_{as} = 1$ and

$$\begin{aligned} ARE_{GC,AO} &= \frac{E^2(g,f)}{I(f)} \\ &= \frac{\tilde{\xi}_{GC}}{I(f)} \end{aligned} \quad (2\text{-}74)$$

In general $\tilde{\xi}_{GC}$ is the *normalized* efficacy

$$\begin{aligned} \tilde{\xi}_{GC} &= \frac{\xi_{GC}}{P_s^2} \\ &= \rho_{as}^2 \, E^2(g,f) \end{aligned} \quad (2\text{-}75)$$

We may extend this result quite easily to obtain the ARE of two GC detectors, D_{GC1} and D_{GC2}, based on nonlinearities g_1 and g_2 and the same coefficients $a_i = s_i$ as

$$ARE_{GC1,GC2} = \frac{\tilde{\xi}_{GC1}}{\tilde{\xi}_{GC2}} \quad (2\text{-}76)$$

where

$$\tilde{\xi}_{GC\,i} = \frac{\left[\int\limits_{-\infty}^{\infty} g_i(x) f'(x)\, dx\right]^2}{\int\limits_{-\infty}^{\infty} g_i^2(x) f(x)\, dx} \quad (2\text{-}77)$$

Notice that the above result for the normalized efficacy of any GC detector could have been derived formally by applying the Pitman-Noether result of (1-41).

In our development above we considered sequences of GC detectors obtained with fixed characteristics g for different sample sizes n. More generally, we can consider sequences of GC detectors with characteristics $g(x,n)$ which are a function of n. One such case is that of (2-55). The quantity $E(g,f)$ is again the asymptotic ratio of the mean to the standard deviation of the GC test statistic under $K_{1,n}$ when $n \to \infty$ (assuming asymptotic normality). Stated loosely, if $g(x,n)$ converges to some fixed $\tilde{g}(x)$, then $E(g,f)$ becomes $E(\tilde{g},f)$. What we really need is convergence of the mean and variance of $g(X_i,n)$ as required under the conditions for Theorem 3 in Section 1.6.

Our results indicate that if our sequence of GC detectors is an AO sequence, than the $ARE_{GC,AO} = 1$. Conversely, if $ARE_{GC,AO} = 1$ and $\tilde{\xi}_{GC} = I(f)$, we find that the sequence of GC detectors is an AO sequence. One conclusion we have drawn from (2-70) and (2-73) is that the efficacy of a detector indicates the *absolute* power level obtainable from it in the asymptotic case. Modifying (2-70) slightly, we get, with σ^2 the noise variance,

$$p_d(\theta \mid D_{GC,n}) \approx 1 - \Phi\left[\Phi^{-1}(1-\alpha) - \frac{\theta}{\sigma}\sqrt{n}\ P_s \rho_{as} \sigma E(g,f)\right] \quad (2\text{-}78)$$

from which we deduce that it is really the quantity $\sigma^2 \xi_{GC}$ which gives an indication of the large sample-size power of the detector for any given value of signal-to-noise ratio (SNR) θ/σ. For the LO and AO detectors we have that $\sigma^2 \xi_{GC} = P_s^2 \sigma^2 I(f)$; thus it is of interest to find the worst-case density minimizing $\sigma^2 I(f)$. It can be shown (Problem 2.4) that the density function minimizing $\sigma^2 I(f)$ is just the Gaussian density function. Thus we may conclude that with optimal processing it is always possible to obtain better performance for non-Gaussian noise than with Gaussian noise under the same SNR conditions. In some cases the potential improvements in performance are quite large.

As an example of the application of these results, consider the LC detector for which $g(x) = x$. From (2-77) we find that the normalized efficacy of the LC detector is $\tilde{\xi}_{LC} = 1/\sigma^2$, where σ^2 is the noise variance. For the sign correlator detector for which $g(x) = \text{sgn}(x)$, we have the normalized efficacy

$$\tilde{\xi}_{SC} = \int_{-\infty}^{\infty} \text{sgn}(x) f'(x)\, dx$$

$$= 4f^2(0)\ . \quad (2\text{-}79)$$

Thus we have

$$ARE_{SC,LC} = 4\sigma^2 f^2(0) \quad (2\text{-}80)$$

This has a value of 0.64 for Gaussian noise and a value of 2.0 for noise with the double-exponential pdf.

In concluding this discussion of asymptotic performance, let us observe that the efficacy may be interpreted as an *output SNR* of a detector. Defining the output SNR of a GC detector as

$$SNR_0 = \frac{[E\{T_{GC}(\mathbf{X}) \mid K_1\} - E\{T_{GC}(\mathbf{X}) \mid H_1\}]^2}{V\{T_{GC}(\mathbf{X}) \mid K_1\}} \quad (2\text{-}81)$$

let us consider the case where θ is small. Then it is easily seen that SNR_0 is approximately $n\theta^2\hat{\rho}_{ae}^2\hat{P}_e^2 E^2(g,f)$, where $\hat{\rho}_{ae}$ and $\hat{P}_e$ are the finite-n versions of the limiting values ρ_{ae} and P_e. The quantity $\hat{\rho}_{ae}^2\hat{P}_e^2 E^2(g,f)$ is called the *differential SNR* or DSNR, and is essentially the same as the efficacy. A general discussion of second-order measures such as these may be found in [Gardner, 1980].

2.4.2 Finite-Sample-Size Performance

Suppose we wish to compare the exact performances of two GC detectors operating in noise with a given probability density function. Then, as we have noted at the beginning of this section, we need in general to consider all combinations of sample size n, signal-to-noise ratio (or signal amplitude θ), and false-alarm probability p_f, and compute the associated detection probabilities. The detection probabilities can, in principle, always be computed using numerical techniques; in practice this may be costly and tedious. Even after all these computations have been made, no *single* index of relative performance can be given, although specific aspects of the relative performance (e.g., relative amplitudes at given n and p_f, and $p_d = 0.9$) may be so characterized. Of course, the entire procedure has to be repeated for a different noise density function.

While the speed and power of modern computing facilities may make such exhaustive comparisons quite feasible, this type of comparison does not yield simple closed-form expressions for indices of relative performance. The availability of such an index is of prime importance as a basis for initial *design* of a good detection scheme. For example, we will see in Chapter 4 how the efficacy can be used as a basis for optimum quantizer design. The efficacy and the ARE are asymptotic absolute performance and relative performance measures, respectively, which are very convenient to use. What is therefore very important to know is the extent to which the ARE of two detectors is a good indicator of their performance under finite-sample-size and non-zero-signal-strength conditions.

The answer to this will obviously depend on the particulars of the two detectors, that is, on their characteristic nonlinearities g; on the noise density function f; and on the combination of sample size n, signal amplitude θ, and false-alarm and detection probabilities p_f and p_d. Note that in using the efficacy and the ARE we make two key assumptions about our detection problem. One is that the sample size n is large enough so that the test statistic of a GC detector may be assumed to have its asymptotic Gaussian distribution. The other is that the signal amplitude θ is small enough to allow us to use the first-order approximations leading to (2-62). What we would like to have, then, are results which allow us to make some general conclusions about the validity of these assumptions in finite-sample-size performance comparisons under various conditions.

Within the class of GC detectors the linear correlator detector is widely used as the optimum detector for Gaussian noise. The second most popular detector in this class is the sign correlator (or sign detector), which is the AO detector for noise with the double-exponential noise density function. This simple detector has excellent asymptotic performance characteristics for heavy-tailed noise density functions, of which the double-exponential noise density function is usually taken as a simple representative. Thus it is natural that the relative performances of the LC and SC detectors for Gaussian and double-exponential noise densities are special cases of considerable interest. It turns out that the ARE generally gives a good indication of finite-sample-size relative performance of these detectors for Gaussian noise, but it may considerably over-estimate the advantage of the sign detector in noise with a double-exponential noise density function. We will see that this is primarily caused by the second key assumption in applying the ARE here: that signal amplitude is low enough to allow use of first-order approximations. This is a rather stringent condition and requires very weak signals (and therefore large sample sizes) for noise densities with the "peaked" behavior at the origin when used with the sign detector, whose characteristic nonlinearity g rises by a single step at the origin. In the case of sign detection of a weak signal, behavior of the noise pdf at the origin is of particular significance, so that the weak-signal approximation is more readily met when the noise density function is flatter at the origin, as in the case of Gaussian noise.

The Sign and Linear Detectors

For the sign correlator detector, in the special case of detection of a *constant* signal for which the SC detector coefficients are all unity, it is relatively easy to compute the finite-sample-size performance in additive i.i.d. noise. This is because the test statistic then has a distribution which is always based on a binomial distribution with parameters (n, q). The value of q is known under the null hypothesis (it is 1/2 for zero-median noise), and with signal present q can be readily determined as $q = P\{W_i > -\theta\}$, which is $1 - F(-\theta)$, where F is the distribution function of the noise.

To compare the finite-sample-size performance of the sign detector with that of another detector we still need to obtain the finite-sample-size performance of the other detector, and this may not be as straightforward. However, in the important special case of comparison of the sign detector with the linear detector for an additive constant signal in Gaussian noise, we are able to perform the comparison with relative ease. The linear detector is simply the linear correlator detector in which the coefficients a_i are again all unity, which is appropriate for the case of additive constant-signal detection. For Gaussian noise the LC detector test statistic has a Gaussian distribution for all sample sizes, making it very easy to compute detection probabilities. This particular finite-sample-size comparison, the sign detector versus the linear detector for constant signal in Gaussian noise, is of considerable interest. The performance of the widely used linear detector (and more generally, of the LC detector) is the benchmark for this type of detection problem, and performance under Gaussian noise is considered important in evaluating detector performance under the signal-present hypothesis. On the other hand, the sign detector is a *nonparametric* detector of a very simple type, and is able to maintain its design value of false-alarm probability for all noise pdf's with the same value of $F(0)$ as the nominal value (e.g., $F(0) = 1/2$ for zero-median noise). Although one may choose to use detectors such as the sign detector for reasons other than to obtain optimum performance in Gaussian noise, performance under conditions of Gaussian noise, a commonly made if not always valid assumption, is required to be adequate.

Figure 2.7 shows the results of a computation of the power functions for constant-signal detection in additive zero-mean Gaussian noise for the sign and linear detectors. Two different combinations of the sample size ($n = 50, 100$) and false-alarm probability ($\alpha = 10^{-2}, 10^{-3}$) were used to get the two plots. We know that the ARE of the sign detector relative to the linear detector for this situation is 0.64. Do these power function plots allow us to make a quantitative comparison with this asymptotic value of 0.64? We have noted earlier that the ARE can be directly related to the asymptotic SNR *ratio* required to maintain the same detection probability, for the same n and α, for two detectors. The SNR ratio at any given detection probability value is influenced by the *horizontal* displacement of the power curves for the sign and linear detectors. A very approximate analysis based on the power functions for $n = 100$ and $\alpha = 10^{-2}$ shows that at both $p_d = 0.8$ and $p_d = 0.4$, we get an SNR ratio (linear detector to sign detector) of approximately 0.63. The SNR here is defined as θ^2/σ^2. This should be compared with the ARE of 0.64 for this situation. A similar rough check at $p_d = 0.4$ and 0.6 for $n = 50$, $\alpha = 10^{-3}$ reveals a slightly smaller SNR ratio. A very preliminary conclusion one may make from such results is that for the values of n and α we have used, the ARE gives a good quantitative indication of actual finite-sample-size relative performance.

This last conclusion (for Gaussian noise) is borne out by some recent results of a study on the convergence of relative efficiency to the ARE done by Michalsky, Wise, and Poor [1982]. In this study the sign and linear detectors have been compared for three types of noise densities. In addition to the Gaussian and double-exponential noise density the *hyperbolic secant* (sech) noise density function defined by

$$f(x) = \frac{1}{2} \operatorname{sech}(\pi x /2) \tag{2-82}$$

was considered. This density function has tail behavior similar to that of the double-exponential function, and has behavior near the origin which is similar to that of the Gaussian density function. These properties can readily be established by considering the behavior of the sech function for large and small arguments. Michalsky et al. present their results as a set of plots showing how the efficiency of the sign detector relative to the linear detector behaves for increasing sample sizes. The relative efficiency is defined as a ratio of sample sizes k and $n(k)$ required by the linear and sign detectors, respectively, to achieve the same values for α and p_d. Their procedure was to pick a value for α and set $p_d = 1 - \alpha$, and then to determine the signal amplitude $\theta(k)$ as a function of k required for the linear detector to achieve this performance, for increasing values of k starting from $k = 1$. Then for each k and signal amplitude $\theta(k)$ the number of samples $n(k)$ required by the sign detector for the same α and p_d was computed. For the double-exponential and sech noise pdf's closed-form expressions are available for the density function of the linear detector test statistic; numerical integration was used to perform the computations for the threshold and randomization probabilities and the detection probabilities.

Figure 2.8 is reproduced from the paper by Michalsky et al. It shows the ratio $k/n(k)$, the computed relative efficiency of the sign detector relative to the linear detector, as a function of k for Gaussian noise and $\alpha = 10^{-4}$. Also shown on this figure is a plot of the estimated relative efficiencies based on use of the Gaussian approximation for the sign detector test statistic. We see that for this case where $\alpha = 10^{-4}$ and $p_d = 1 - 10^{-4}$, the relative efficiency is above 0.55, within about 15% of the ARE, for $k \geq 35$. A general conclusion which can be made from the results obtained for this case is that as α decreases, the relative efficiencies converge more slowly to the ARE value. One reason for this may be that the ARE result is based on Gaussian approximations to the distributions of the test statistics (in this case the sign detector statistic). For decreasing α, the threshold setting obtained from the tail region of the Gaussian approximation will be accurate only for larger sample sizes. Another reason is that in this study p_d was set to equal $1 - \alpha$, so that p_d increases as α decreases, requiring larger signal amplitudes. Before we discuss this in more detail let us note some other interesting results obtained in the paper by Michalsky et al.

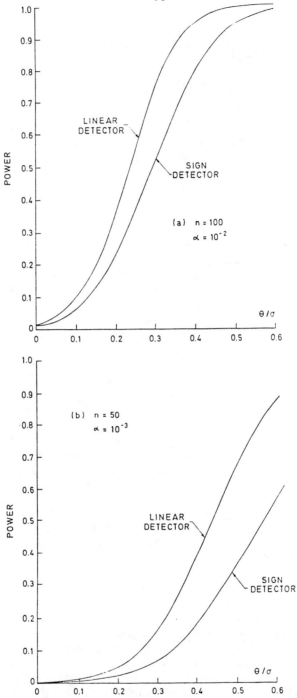

Figure 2.7 Finite-Sample-Size Performance of Linear and Sign Detectors for Constant Signal θ in Additive Gaussian Noise with Variance σ^2

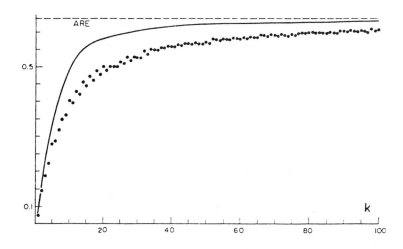

Figure 2.8 Relative Efficiency of the Sign Detector Compared to the Linear Detector for Gaussian Noise at $\alpha = 10^{-4}$. The Smooth Curve Is Obtained from the Gaussian Approximation. (From [Michalsky et al. 1982])

Figures 2.9(a) and (b) and Figure 2.10 are also reproduced from this paper. Figure 2.9(a) for $\alpha = 10^{-3}$ and double-exponential noise shows a very slow convergence of the relative efficiency of the sign detector relative to the linear detector, to the ARE value of 2.0. It also shows that the relative efficiency can be approximated very well by its estimate based on the Gaussian approximation, with the same means and variances as the exact distributions of the sign and linear detector test statistics. The Gaussian approximation is used in the following way: for each value of k the linear detector test statistic is assumed to be Gaussian with mean $k\theta(k)$ and variance $k\sigma^2$. The threshold is set for $\theta(k) = 0$ using the Gaussian distribution, and the signal strength $\theta(k)$ is then obtained for p_d to equal $1 - \alpha$, again using the Gaussian approximation for the linear detector statistic. This signal strength is now an estimate $\hat{\theta}(k)$ of the correct signal strength. This estimated signal strength is then used to find the binomial parameter $q = P\{W_i > -\theta\}$ needed to find the mean and variance of the sign detector statistic. Finally, again based on the Gaussian approximations for the sign detector test-statistic distributions, the sign detector sample size $n(k)$ is determined to allow it to have the same value for α and $p_d = 1 - \alpha$, with the same signal strength $\hat{\theta}(k)$. Use of the Gaussian approximation allows relative efficiencies to be easily computed for larger values of k. The result is shown in Figure 2.9(b), where the relative efficiency estimates are shown as a function of both k and $\hat{\theta}(k)$ for values of k beyond the largest shown in Figure 2.9(a). Figure 2.9(b) shows that to obtain a relative efficiency of 1.5 one needs to use more than 600 observations for the linear detector (and more than 400 observations for the sign detector). This rather slow convergence to the ARE value is our first indication that the use of the ARE may not always be justified when one is interested in the finite-sample-size relative performance of two detectors.

The rather different behaviors of the relative efficiencies of the sign and linear detectors for these two noise densities (Gaussian and double-exponential) is not something which could not have been predicted. We have remarked that we make use of two approximations in obtaining the ARE. One is the Gaussian approximation for the distribution of the test statistics. Figure 2.9(a) shows that this is probably quite valid for this particular case. The other approximation we make is that θ is small enough to essentially allow use of the expansion $1 - F(-\theta) \approx 0.5 + \theta f(0)$, where f is the noise density function. For the case of Gaussian noise the density is smooth at the origin and $f'(0) = 0$. In contrast, the double-exponential density is peaked at the origin with non-zero values for the one-sided derivatives there. For the unit-variance double-exponential density we have $f(0) = 1/\sqrt{2}$ and $f'(0^-) = 1$. Thus the signal has to be weaker before the approximation $1 - F(-\theta) \approx 0.5 + \theta f(0)$ becomes valid for the double-exponential density, as compared to the Gaussian density. Consider, for example, the case corresponding to $k = 100$ in Figure 2.9(b) (for $\alpha = 10^{-3}$), for which the signal strength $\theta(k)$ is estimated to be 0.62. [This is actually $\hat{\theta}(k)/\sigma$, the variance σ^2 being 1 here.] Let us compare the exact value of $1 - F(-\theta)$ to $0.5 + \theta f(0)$ for $\theta = 0.62$ for the unit variance double-exponential density. A simple computation shows that $1 - F(-\theta) = 0.792$, whereas $0.5 + \theta f(0) = 0.938$, a rather poor approximation. On the other hand, for unit-variance Gaussian noise we get $1 - F(-\theta) = 0.732$ and $0.5 + \theta f(0) = 0.747$ for $\theta = 0.62$, which shows that the approximation is very good for Gaussian noise. For $k = 4000$ we have $\hat{\theta}(k) = 0.1$ from Figure 2.9(b). For $\theta = 0.1$ we find for the unit-variance double-exponential density that $1 - F(-\theta) = 0.566$ and $0.5 + \theta f(0) = 0.571$, which are close. The relative efficiency at $k = 4000$ is seen to be approximately 1.75.

In general we would expect that for any noise density with a local maximum at the origin the relative efficiency of the sign detector relative to the linear detector will be less than its ARE, assuming that k is large enough for the Gaussian approximation to be valid. This is because the approximation $1 - F(-\theta) \approx 0.5 + \theta f(0)$ is then always larger than the correct value of $1 - F(-\theta)$, making the ARE a more optimistic measure of

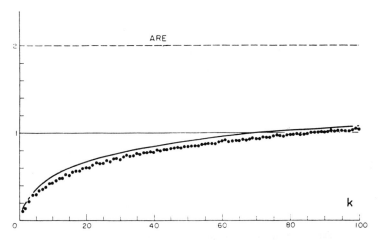

Figure 2.9(a) Relative Efficiency of the Sign Detector Compared to the Linear Detector for Double-Exponential Noise at $\alpha = 10^{-3}$. The Smooth Curve Is Obtained from the Gaussian Approximation. (From [Michalsky et al. 1982])

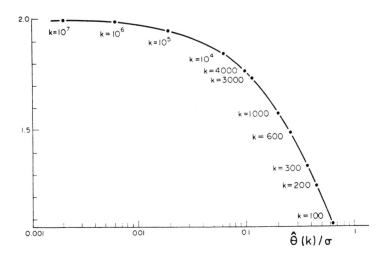

Figure 2.9(b) The Gaussian Approximation to the Relative Efficiency of the Sign Detector Compared to the Linear Detector for Double-Exponential Noise at $\alpha = 10^{-3}$, as a Function of the Gaussian Approximation $\hat{\theta}(k)$ to the Signal Strength (Normalized by σ) (From [Michalsky et al. 1982])

the sign detector relative performance. Since in the study by Michalsky et al. [1982] it was always assumed for computational convenience that $p_d = 1 - \alpha$, decreasing α leads to increasing p_d and hence increasing signal strengths for each value of k. Thus it is to be expected that the relative efficiencies will converge more slowly to the ARE as α is decreased.

Figure 2.10 shows a set of plots for different values of α of the relative efficiency of the sign detector relative to the linear detector for the sech noise density function. The ARE value here is 1. The slower convergence rate as α is decreased is apparent in this figure. In addition, we see the expected convergence behavior, which is between that for the double-exponential and Gaussian densities. Prior to the above-described study Miller and Thomas [1975] had also undertaken a numerical study on the convergence of the relative efficiency to the ARE of an asymptotically optimum sequence of detectors compared to the linear detector for the detection of a constant signal in noise with a double-exponential density function. The AO sequence of detectors used was the sequence of GC detectors which are Neyman-Pearson optimum for each n; with increasing n the signal amplitude decreases to maintain a fixed value for the detection probability. The relative efficiency in this numerical study was defined as a ratio of sample sizes required to make the performances of two detectors the same at given values of $p_f = \alpha$ and p_d, for a given value of signal amplitude. The conclusion of Miller and Thomas was also that in this particular situation the relative efficiency can converge rather slowly to its maximum value, which is the ARE value of 2.0. Some further numerical results for this particular case of detection of a constant signal in noise with a double-exponential density function have been given by Marks, Wise, Haldeman, and Whited [1978]. The case of time-varying signal has been considered by Liu and Wise [1983]. More recently Dadi and Marks [1987] have studied further the relative performances of the linear, sign, and optimum detectors for double-exponential noise pdf's and have again reached the same conclusion.

Another type of GC detector which is not too difficult to analyze for finite-sample-size performance is the quantizer-correlator (QC) detector which we shall consider in Chapter 4. In the QC detector the characteristic g of a GC detector is a piecewise-constant quantizer nonlinearity q. For Gaussian noise the performances of QC detectors relative to the LC detector generally follow quite closely the predictions based on asymptotic comparisons. For the double-exponential noise density Michalsky et al. [1982] also considered the simple three-level symmetric QC detector with a middle level of zero, called the *dead-zone limiter* (DZL) detector. Their numerical results indicate that the efficiency of the DZL detector relative to the linear detector for this noise density does converge more rapidly to the ARE value

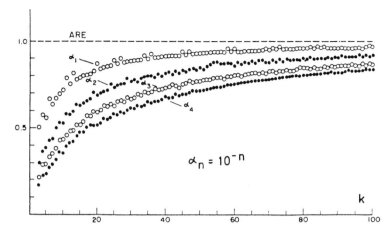

Figure 2.10 Relative Efficiency Curves for the Sign Detector Compared to the Linear Detector for Sech Noise, for Different Values of α (From [Michalsky et al. 1982])

than is the case for the sign detector relative to the linear detector. In fact here one observes very interesting behavior; the relative efficiency actually overshoots the ARE value and then converges to it from above. It is possible to give this overshoot phenomenon an explanation similar to that which explains the convergence from below to the ARE values in our discussion above.

In several other sections in the following chapters we will discuss specific finite-sample-size performance results for particular detectors. Our general conclusion may be summarized as follows. In most cases the use of efficacies and ARE's allows one to make valid qualitative comparisons of different detectors. In many cases the ARE gives a fairly good quantitative indication of finite-sample-size relative performance. In some cases, particularly for the sign detector and pdf's such as the double-exponential noise density function, convergence to the asymptotic value may be quite slow. As a basis for the preliminary design, analysis, and comparison of detection systems, the use of efficacies and ARE's are very convenient and appropriate. As a second step it is highly desirable that some exact finite sample detection performance characteristics be obtained, or that the detector be simulated, to back up conclusions of asymptotic analysis, to fine-tune a design, or to make clear the ranges of applicability of the ARE in performance comparisons. We have also seen that sometimes a simple analysis of the asymptotic approximations made in defining the ARE can give some valuable insights about finite-sample-size relative performance.

2.5 Locally Optimum Bayes Detection

We will now briefly consider the binary signaling problem where one wishes to decide between two known signal sequences in the model given by (2-1), rather than make a choice between noise only and a known signal in noise for the X_i.

Let $\{s_{0i}, i = 1,2,...,n\}$ and $\{s_{1i}, i = 1,2,...,n\}$ be two known signal sequences of which one is present in the observations, with non-zero amplitude θ. Retaining our assumption of independence and identical distributions for the additive noise components W_i, the likelihood ratio for this hypothesis-testing problem becomes

$$L(\mathbf{X}) = \prod_{i=1}^{n} \frac{f(X_i - \theta s_{1i})}{f(X_i - \theta s_{0i})} \tag{2-83}$$

Let $p_j > 0$ be the *a priori* probability that the signal sequence $\{s_{ji}, i = 1,2,...,n\}$ has been received in noise, $j = 0,1$. Of course $p_0 + p_1 = 1$. Let c_{jk} be the cost of deciding that the j-th signal was transmitted when in fact the k-th one was transmitted, for $j,k = 0,1$. We will assume that $c_{jj} = 0$ and that $c_{jk} > 0$ for $j \neq k$. Then the Bayes detector minimizing the Bayes risk is one implementing the test

$$L(\mathbf{X}) > \frac{c_{10} p_0}{c_{01} p_1} \tag{2-84}$$

With $L(\mathbf{X})$ exceeding the right-hand side above the signal sequence $\{s_{1i}, i = 1,2,...,n\}$ is decided upon.

Denoting by $\exp(k)$ the right-hand side of (2-84), we have the equivalent form for the above test,

$$\sum_{i=1}^{n} \ln f(X_i - \theta s_{1i}) > k + \sum_{i=1}^{n} \ln f(X_i - \theta s_{0i}) \tag{2-85}$$

Now using a Taylor series expansion for $\ln f(x - u)$ about $u = 0$, assuming sufficient regularity conditions on f, we obtain as another equivalent form of the above test,

$$\theta \sum_{i=1}^{n} (s_{1i} - s_{0i}) g_{LO}(X_i)$$
$$+ \frac{\theta^2}{2} \sum_{i=1}^{n} (s_{1i}^2 - s_{0i}^2) \left[\frac{f''(X_i)}{f(X_i)} - g_{LO}^2(X_i) \right] + o(\theta^2) > k \tag{2-86}$$

where g_{LO} has been defined in (2-33).

This result allows some interesting interpretations. For a finite sample size n, in the case of vanishing signal strengths (the local case), by ignoring second-order and higher-order terms we find that the Bayes detector implements the locally optimum Bayes test

$$\theta \sum_{i=1}^{n} (s_{1i} - s_{0i}) g_{LO}(X_i) > k \qquad (2\text{-}87)$$

Notice that the test statistic on the left-hand side above contains the signal amplitude θ explicitly. Suppose, for example, that k is positive because $c_{10} > c_{01}$ and $p_0 = p_1$ in (2-84). This means that the error of falsely deciding that the s_{1i}, $i = 1,2,...,n$, were transmitted is more costly than the other type of error. Then for a *given* set of observations $X_1, X_2, ..., X_n$, the assumed small value of $\theta \ll \sigma$ will influence the final decision made by the detector. As θ becomes smaller the observations become less reliable as an objective means for discriminating between the two signal alternatives, and the decision which is less costly when no observations are available tends to be made by the detector.

In the special but commonly assumed case of equal costs and prior probabilities we have $k = 0$ in (2-87) and θ can then be dropped from the left-hand side. The resulting locally optimum Bayes detector than has the structure of the LO detector based on the Neyman-Pearson criterion, with the threshold now fixed to be zero.

In considering the asymptotic situation $n \to \infty$ and $\theta \to 0$ simultaneously, we have to specify the relationship between θ and n. Let us assume, as we have done in Section 2.3, that $\theta = \gamma/\sqrt{n}$ for some fixed $\gamma > 0$. Further, we will assume that the signal sequences have finite average powers,

$$0 < \lim_{n \to \infty} \frac{1}{n} \sum_{i=1}^{n} s_{ji}^2 = P_j^2 < \infty \qquad (2\text{-}88)$$

for $j = 0,1$. Under these conditions we find that the second plus higher-order term on the left-hand side of (2-86) converges in probability under both alternatives to a constant which is

$$\lim_{n \to \infty} \frac{\gamma^2}{2n} \sum_{i=1}^{n} (s_{1i}^2 - s_{0i}^2) \left[\frac{f''(X_i)}{f(X_i)} - g_{LO}^2(X_i) \right]$$

$$= -\frac{\gamma^2}{2} (P_1^2 - P_0^2) I(f) \qquad (2\text{-}89)$$

Here $I(f)$ is the Fisher information for location shift for the noise pdf f.

Using this result we obtain the locally optimum Bayes test, which retains its optimality under the specific asymptotic situation $n \to \infty$, $\theta \to 0$, and $\theta\sqrt{n} \to \gamma > 0$ to be

$$\theta \sum_{i=1}^{n} (s_{1i} - s_{0i}) g_{LO}(X_i) > k + \frac{\gamma^2}{2}(P_1^2 - P_0^2) I(f) \quad (2\text{-}90)$$

The effect of letting θ approach zero for a fixed, finite n is to make $\gamma \to 0$ (set $\gamma = \theta\sqrt{n}$), which gives us the locally optimum Bayes detector. We find that when $P_1^2 = P_0^2$ the locally optimum Bayes detector employing the test of (2-87) does retain its optimality asymptotically. Otherwise, the second "bias" term has to be added to the threshold in (2-90) as a necessary condition for asymptotic optimality.

The important practical result from the above development is that for the commonly assumed case of equal costs for the two types of errors, equal *a priori* probabilities for the two alternatives about the signal, and antipodal signaling ($s_{1i} = -s_{0i}$, $i = 1,2,...,n$), the locally optimum Bayes detector is based on the test

$$\sum_{i=1}^{n} (s_{1i} - s_{0i}) g_{LO}(X_i) > 0 \qquad (2\text{-}91)$$

This also yields an asymptotically optimum sequence of Bayes detectors. More generally a non-zero threshold, modified by an additive bias term, and normalized by dividing with the signal amplitude θ, is required.

Our results for the locally and asymptotically optimum Bayes detectors required particular attention to be focused on the "bias" term in the threshold and on the signal amplitude θ which is in general necessary to obtain the correct threshold normalization. This is to be contrasted with the LO and AO detectors based on the Neyman-Pearson criterion. There the threshold setting is decided at the end to obtain the required false-alarm probability, any simplifying monotone transformations of the test statistic being permissible in the intermediate steps in deriving the final structure of the threshold test.

The derivation and analysis of locally and asymptotically optimum Bayes detection schemes for known binary signals has been considered by Middleton [1966] and also more recently by him [1984]. This latter paper, as well as Spaulding and Middleton [1977a], give detection performance results based on upper bounds on the error probabilities and also based on the asymptotic Gaussian distributions of the test statistic, for noise pdf's modeling impulsive noise. Some performance results have also been given by Spaulding [1985] for the LO Bayes detector. Maras, Davidson, and Holt [1985a] give detection performance results for the Middleton Class A noise model which we shall discuss briefly in the next chapter. An interesting sub-optimum GC detection scheme has also been suggested by Hug [1980].

2.6 Locally Optimum Multivariate Detectors

In this last section we will consider a generalization of our basic model (2-1) for the univariate observations X_i that we have been concerned with so far. We will here study the case where we have a set of i.i.d. *multivariate* or vector observations $\mathbf{Y}_i = (X_{1i}, X_{2i}, ..., X_{Li})$ described by

$$\mathbf{Y}_i = \theta \mathbf{r}_i + \mathbf{V}_i, \quad i = 1,2,...,n \tag{2-92}$$

Now the $\mathbf{r}_i = (s_{1i}, s_{2i}, \ldots, s_{Li})$ are known L-variate signal vectors and the $\mathbf{V}_i = (W_{1i}, W_{2i}, \ldots, W_{Li})$ are, for $i = 1,2,...,n$, a set of i.i.d. random noise vectors with a common L-variate density function f.

Such a model for multivariate observations can describe the outputs of L receivers or sensors forming an array, designed to pick up a signal from a distant source in a background of additive noise. At each sampling time the L sensor outputs are obtained simultaneously. The noise in each sensor may be correlated with noise in other sensors at any one sampling instant, although our model above requires the noise samples to be temporally independent. The signal arrives as a propagating plane wave from a particular direction, so that the components of each $\mathbf{r}_i$ are delayed versions of some common signal waveform in this type of problem.

Another application in which the model is appropriate is that in which a scalar observed waveform possibly containing a known signal is sampled in periodic bursts or groups of L closely spaced samples. If sufficient time is allowed between each group of L samples the groups may be assumed to be independent, although the samples within each group may be dependent. Such a scheme can be used to improve the performance of a detector which is constrained to operate on independent samples only, allowing it to use more data in the form of independent multivariate samples. We shall see in Chapter 5 that for detection of a narrowband signal which is completely known in additive narrowband noise we can interpret the in-phase and quadrature observation components as having arisen from such a sampling scheme with $L = 2$.

Under the signal-present hypothesis the density function f_Y of the set of n observation vectors $\mathbf{Y} = (\mathbf{Y}_1, \mathbf{Y}_2, ..., \mathbf{Y}_n)$ is now

$$f_Y(\mathbf{y} \mid \theta) = \prod_{i=1}^{n} f(\mathbf{y}_i - \theta \mathbf{r}_i) \tag{2-93}$$

keeping in mind that f is now a function of an L-component vector quantity. The LO detector now has a test statistic

$$\lambda_{LO}(\mathbf{Y}) = \frac{\left.\frac{d}{d\theta} f_Y(\mathbf{Y} \mid \theta)\right|_{\theta=0}}{f_Y(\mathbf{Y} \mid 0)}$$

$$= \sum_{i=1}^{n} \frac{\left.\frac{d}{d\theta} f(\mathbf{Y}_i - \theta \mathbf{r}_i)\right|_{\theta=0}}{f(\mathbf{Y}_i)}$$

$$= \sum_{i=1}^{n} \mathbf{r}_i \left[\frac{-\nabla f(\mathbf{Y}_i)}{f(\mathbf{Y}_i)}\right]^T \tag{2-94}$$

where ∇f is the gradient vector defined by

$$\nabla f(\mathbf{v}) = \left[\frac{\partial f(\mathbf{v})}{\partial v_1}, \frac{df(\mathbf{v})}{\partial v_2}, \ldots, \frac{\partial f(\mathbf{v})}{\partial v_L}\right] \tag{2-95}$$

This result reduces to our result of (2-32) for $L = 1$. In our more general case here the locally optimum vector nonlinearity is

$$\mathbf{g}_{LO}(\mathbf{y}_i) = -\frac{\nabla f(\mathbf{y}_i)}{f(\mathbf{y}_i)}$$

$$= -\nabla \ln f(\mathbf{y}_i) \tag{2-96}$$

We see that if $f(\mathbf{v}) = \prod_{j=1}^{L} f(v_j)$ the test statistic of (2-94) is simply our earlier LO statistic of (2-32), applied to nL independent observations.

Consider a multivariate GC test statistic

$$T_{GC}(\mathbf{Y}) = \sum_{i=1}^{n} \mathbf{a}_i \mathbf{g}^T(\mathbf{Y}_i) \tag{2-97}$$

where $\mathbf{g}(\mathbf{y}_i) = [g_1(\mathbf{y}_i), g_2(\mathbf{y}_i), ..., g_L(\mathbf{y}_i)]$, each g_j being a real-valued function of an L-vector. It can be shown quite easily that the efficacy of such a test statistic for our detection problem is

$$\xi_{GC} = \lim_{n \to \infty} \frac{\left[\sum_{i=1}^{n} \mathbf{a}_i \int_{R^L} \mathbf{g}^T(\mathbf{v}) \mathbf{g}_{LO}(\mathbf{v}) f(\mathbf{v}) \, d\mathbf{v} \, \mathbf{r}_i^T \right]^2}{n \sum_{i=1}^{n} \mathbf{a}_i \int_{R^L} \mathbf{g}^T(\mathbf{v}) \mathbf{g}(\mathbf{v}) f(\mathbf{v}) \, d\mathbf{v} \, \mathbf{a}_i^T}. \qquad (2\text{-}98)$$

With $\mathbf{g} = \mathbf{g}_{LO}$ and $\mathbf{a}_i = \mathbf{r}_i$ we get the maximum efficacy

$$\xi_{GC,LO} = \lim_{n \to \infty} \frac{1}{n} \sum_{i=1}^{n} \mathbf{r}_i \, \mathbf{I}(f) \mathbf{r}_i^T \qquad (2\text{-}99)$$

where $\mathbf{I}(f)$ is the matrix defined by

$$\mathbf{I}(f) = E\{\mathbf{g}_{LO}^T(\mathbf{V}_i) \mathbf{g}_{LO}(\mathbf{V}_i)\} \qquad (2\text{-}100)$$

which is Fisher's information matrix corresponding to the function $I(f)$ of (2-12) in the univariate case.

Suppose that f is the multivariate Gaussian density,

$$f(\mathbf{v}) = \frac{1}{(2\pi)^{L/2} (\det \Lambda)^{1/2}} e^{-\mathbf{v} \Lambda^{-1} \mathbf{v}^T / 2} \qquad (2\text{-}101)$$

where Λ is the covariance matrix. Then we find that

$$\mathbf{g}_{LO}(\mathbf{y}_i) = \mathbf{y}_i \Lambda^{-1} \qquad (2\text{-}102)$$

and the LO test statistic is the matched filter statistic

$$\lambda_{LO}(\mathbf{Y}) = \sum_{i=1}^{n} \mathbf{r}_i \Lambda^{-1} \mathbf{Y}_i^T \qquad (2\text{-}103)$$

In Chapter 5 we will be considering essentially LO test statistics for the bivariate case with the bivariate density function f being *circularly symmetric* so that $f(v_1, v_2) = h(\sqrt{v_1^2 + v_2^2})$. A special case of this is the bivariate Gaussian density for two i.i.d. Gaussian components. The model for multivariate observations that we have treated here has been used by Martinez, Swaszek, and Thomas [1984] to obtain locally optimum detectors.

PROBLEMS

Problem 2.1

Verify that the maximum likelihood estimate $\hat{\theta}_{ML}$ of θ under K_1 of (2-14) for the double-exponential noise pdf is obtained as a solution of (2-24).

Problem 2.2

Draw a block diagram for an implementation of the test statistic $\tilde{\lambda}(\mathbf{X})$ of (2-22) and (2-23), using multipliers, adders, accumulators, and a fixed nonlinearity.

Problem 2.3

Show that the limiting value of the power of a sequence of size α GC detectors for $H_{1,n}$ versus $K_{1,n}$ of (2-42) with $\theta_n = \gamma/\sqrt{n}$ is given by (2-65).

Problem 2.4

Prove that the Gaussian pdf minimizes $\sigma^2 I(f)$, where $I(f)$ is the Fisher information for location, from among all absolutely continuous pdf's.

Problem 2.5

Consider the GC detector based on the soft-limiter nonlinearity

$$g(x) = \begin{cases} c, & x > c \\ x, & |x| \leq c \\ -c, & x < -c \end{cases}$$

for some constant $c \geq 0$. Obtain the ARE of this GC detector relative to the LC detector (both using the same GC coefficients) for H_1 versus K_1 of (2-13) and (2-14), when the pdf f is symmetric about the origin. Obtain the limiting values of the ARE for $c \to 0$ and $c \to \infty$. Compute the ARE for $c = 3\sigma$ with f the Gaussian density with variance σ^2. Comment on the implication of this result.

Problem 2.6

Consider the GC detector based on the dead-zone limiter

nonlinearity

$$g(x) = \begin{cases} 1, & x > c \\ 0, & |x| \leq c \\ -1, & x < -c \end{cases}$$

for some constant $c \geq 0$. Obtain the ARE of this GC detector relative to the LC detector for symmetric f in H_1 versus K_1 of (2-13) and (2-14). Obtain the limiting value for the ARE when $c \to 0$. Find the optimum value of c maximizing the ARE for Gaussian f, and determine the maximum ARE value.

Problem 2.7

The "hole-puncher" function

$$g(x) = \begin{cases} x, & |x| \leq c \\ 0, & |x| > c \end{cases}$$

may be used as the detector nonlinearity for a GC detector. Obtain the efficacy of the resulting detector for f the Cauchy pdf

$$f(x) = \frac{1}{\pi\sigma} \frac{1}{1 + (x/\sigma)^2}$$

in testing H_1 versus K_1 of (2-13) and (2-14), and find the optimum value of c/σ maximizing this efficacy. What is the limiting value of the efficacy as $c \to \infty$? Explain your result.

Problem 2.8

The characteristic g of a GC detector is a linear combination of functions in a set $\{g_1, g_2, \ldots, g_M\}$ which is orthonormal with respect to the weighting function f, the noise pdf. Show that the efficacy of an AO GC detector of this type is the sum of the efficacies of detectors based on the individual g_i, $i = 1, 2, \ldots, M$.

Problem 2.9

Develop a possible explanation for the overshoot phenomenon mentioned at the end of Section 2.4 in our discussion of numerical results on convergence of relative efficiencies to the ARE for the dead-zone limiter detector relative to the linear detector.

Chapter 3

SOME UNIVARIATE NOISE PROBABILITY DENSITY FUNCTION MODELS

3.1 Introduction

In this chapter we will apply the results we have derived so far to some particular models for univariate non-Gaussian noise probability density functions. The models we will consider have been found to be appropriate for modeling non-Gaussian noise in different systems. A large number of investigations have been carried out over the course of the last forty years on the characteristics of noise processes encountered in different environments. In particular, there has been much interest in characterization of underwater acoustic noise, urban and man-made RF noise, low-frequency atmospheric noise, and radar clutter noise. For such situations it is well-recognized that the mathematically appealing simple Gaussian noise model is often not at all appropriate. It is not within the scope of this book to detail the development of the non-Gaussian noise models that we will describe here, and we will have to refer the reader to the relevant literature to gain an understanding of the basis for the models.

There is one common feature, however, that all these models share. This is that they specify noise density functions which in their tails decay at rates lower than the rate of decay of the Gaussian density tails. The essential implication of this fact is that non-Gaussian noise which tends to occur in practice is generally more likely to produce large-magnitude observations than would be predicted by a Gaussian model (satisfying the same constraint on some scale characteristic, such as a noise percentile). We should then expect the optimum and locally optimum detectors to employ some nonlinear characteristic g tending to reduce the influence of large-magnitude observations on the test statistic. This should be expected because large-magnitude observations will now be less reliable (compared to the Gaussian case) in indicating the absence or presence of a signal. We will, indeed, observe this general characteristic.

As an example illustrating the above statements, let us briefly examine some results which have been reported for the noise affecting ELF (extremely low frequency) communication systems. In studies described by Evans and Griffiths [1974], the non-Gaussian atmospheric noise caused by lightning activity in local and distant storms is identified as the major limiting factor in long-range ELF communication systems operating in the range 3 to 300 Hz. The characteristic feature of noise in this band is the occurrence of impulse-like components, in a background noise waveform due to low attenuation of numerous distant atmospheric noise sources. A typical recording of a noise waveform in this band of frequencies is shown in Figure 3.1. Figure 3.2 shows a typical amplitude probability distribution function for such noise, obtained from empirical data. For comparison, Figure 3.2 also shows a Gaussian distribution function with the same variance. Evans and Griffiths [1974] also present an interesting graph showing the locally optimum nonlinearity g_{LO} obtained for an empirically determined noise probability density function; this is given here as Figure 3.3. Further details on receiver structures employing such nonlinear processing for ELF systems have been given by Bernstein, McNeil, and Richer [1974] and Rowe [1974]. An overview of ELF communication systems may be found in [Bernstein et al., 1974]. Modestino and Sankur [1974] have given some specific models and results for ELF noise. More recently, several suboptimal detectors as well as the locally optimum detector have been studied by Ingram [1984], who obtained detection performance characteristics for ELF atmospheric noise.

We will obtain nonlinear characteristics of the nature of that shown in Figure 3.3 for the types of noise models we will now consider analytically. The first type of model we will consider defines noise density functions as generalizations of well-known univariate densities, one of them the Gaussian density. These have been found to provide good fits in some empirical studies. The second case we will consider is that of mixture noise, which can be given a statistical as well as a physical basis. It will also be shown that a useful noise model developed by Middleton is closely related to the mixture model. We conclude this chapter with a brief discussion of strategies for adaptive detection when the noise probability density function cannot be assumed to be *a priori* known.

3.2 Generalized Gaussian and Generalized Cauchy Noise

The convenient mathematical properties of the Gaussian probability density function have to be given up, at least partially, in order to obtain density function models which are better descriptions of the noise and interference encountered in many real-world systems. As we have remarked above, the presence of impulsive components tends to produce noise density functions with heavier tails than those of the Gaussian densities. One way to obtain such modified tail behavior is to start with the Gaussian density function and to allow its rate of exponential decay to become a free parameter. In another approach we start at the other extreme case of algebraic tail behavior of the Cauchy noise probability density function, and introduce free parameters to allow a range of possibilities which includes the Gaussian as a special case.

3.2.1 Generalized Gaussian Noise

A *generalized Gaussian* noise density function is a symmetric, unimodal density function parameterized by two constants, the variance σ^2 and a rate-of-exponential-decay parameter $k > 0$. It is defined by

$$f_k(x) = \frac{k}{2A(k)\Gamma(1/k)} e^{-[|x|/A(k)]^k} \tag{3-1}$$

where

$$A(k) = \left[\sigma^2 \frac{\Gamma(1/k)}{\Gamma(3/k)}\right]^{1/2} \tag{3-2}$$

and Γ is the gamma function:

$$\Gamma(a) = \int_0^\infty x^{a-1} e^{-x} \, dx \tag{3-3}$$

We find that for $k = 2$ we get the Gaussian density function, and for lower values of k the tails of f_k decay at a lower rate than for the Gaussian case. The value $k = 1$ gives us the double-exponential noise density function. Figure 3.4 shows the shapes of $f_k(x)$ for different k, with the variance $\sigma^2 = 1$ for all k. Thus the class of generalized Gaussian density functions allows us to conveniently consider a spectrum of densities ranging from the Gaussian to those with relatively much faster or much slower rates of exponential decay of their tails. Algazi and Lerner [1964] indicate that generalized Gaussian densities with k around 0.5 can be used to model certain impulsive atmospheric noise.

For the generalized Gaussian noise densities we find that the locally optimum nonlinearities g_{LO} are, from (2-33),

$$g_{LO}(x) = \frac{k}{[A(k)]^k} |x|^{k-1} \text{sgn}(x) \tag{3-4}$$

These are, as expected, odd-symmetric characteristics, since the f_k are even functions. Note that for $k = 2$ we get $g_{LO}(x) = x/\sigma^2$, the result of (2-36), and for $k = 1$ we get $g_{LO}(x) = (\sqrt{2}/\sigma)\text{sgn}(x)$, which agrees with the result of (2-38). Figure 3.5 shows the normalized versions $\{[A(k)]^k/k\}g_{LO}(x)$ for the generalized Gaussian densities. Observe that the value $k = 1$ separates two distinct types of behavior. For $k > 1$ the nonlinearities g_{LO} are continuous at the origin and increase monotonically to $+\infty$ for increasing x, whereas for $k < 1$ there is an infinite discontinuity at the origin and g_{LO} decreases monotonically to zero as x increases from zero.

It is quite easy to establish that the Fisher information $I(f_k)$ for the generalized Gaussian density function is

$$I(f_k) = \frac{k^2 \Gamma(3/k) \Gamma(2 - 1/k)}{\sigma^2 \Gamma^2(1/k)} \qquad (3\text{-}5)$$

which is finite for $k > 0.5$. We can now compare the performance of the LC detector, which is optimum for Gaussian noise, with that of the LO detector based on g_{LO} of (3-4) when the noise density is the generalized Gaussian density f_k. From (2-77) we easily obtain the normalized efficacy $\tilde{\xi}_{LC}$ of the LC detector for any noise density function f to be

$$\tilde{\xi}_{LC} = \frac{1}{\sigma^2} \qquad (3\text{-}6)$$

where σ^2 is the noise variance. Then we find that the ARE of the LO detector for noise pdf f_k, relative to the LC detector, computed for the situation where f_k is the noise density function, is [from (2-76)]

$$\begin{aligned} ARE_{LO,LC} &= 1/ARE_{LC,LO} \\ &= \sigma^2 I(f_k) \\ &= \frac{k^2 \Gamma(3/k) \Gamma(2 - 1/k)}{\Gamma^2(1/k)} \end{aligned} \qquad (3\text{-}7)$$

For $k = 2$ we have, of course, $ARE_{LO,LC} = 1$, and $k = 1$ gives us $ARE_{SC,LC} = 2$ for double-exponential noise. Figure 3.6 shows $ARE_{LO,LC}$ as a function of k. We emphasize that this is the ARE of the LO detector for f_k relative to the LC detector, when f_k is the noise density. Thus explicit knowledge of the non-Gaussian nature of the noise density function can be exploited to get significantly more efficient schemes than the LC detector.

As an example of the comparison of two *fixed* nonlinearities, let us consider $ARE_{SC,LC}$ as a function of k for the generalized Gaussian noise pdf given by (3-1). Applying the result (2-76), we conclude easily that

$$\begin{aligned} ARE_{SC,LC} &= 4\sigma^2 f_k^2(0) \\ &= \frac{k^2 \Gamma(3/k)}{\Gamma^3(1/k)} \end{aligned} \qquad (3\text{-}8)$$

Because the sign correlator (SC) detector is the LO detector for $k = 1$, we find that $ARE_{SC,LC}$ agrees with $ARE_{LO,LC}$ of (3-7) for

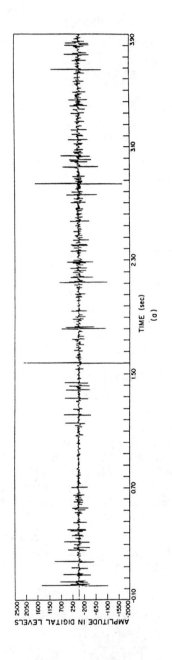

Figure 3.1 Sample Function of Typical ELF Noise
Process (from Evans and Griffiths [1974])
© 1974 IEEE

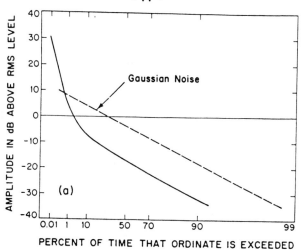

Figure 3.2 Amplitude Probability Distribution of Typical ELF Noise (from Evans and Griffiths [1974])
© 1974 IEEE

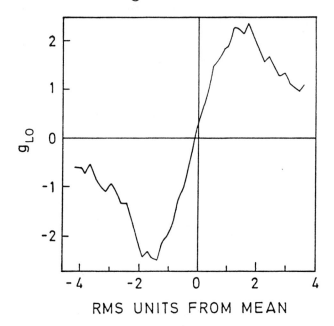

Figure 3.3 Locally Optimum Nonlinearity for Empirically Determined ELF Noise PDF (from Evans and Griffiths [1974]) © 1974 IEEE

$k = 1$. Otherwise, $ARE_{SC,LC}$ is less than $ARE_{LO,LC}$. Nonetheless, as shown in Figure 3.6, the simple SC detector is significantly more efficient that the LC detector for all $k \leq 1$.

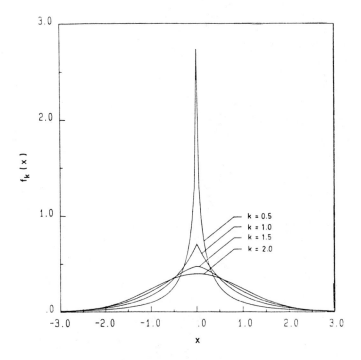

Figure 3.4 Generalized Gaussian Probability Density Functions ($\sigma^2 = 1$)

3.2.2 Generalized Cauchy Noise

The other class of noise density functions which is useful in studying the shapes of the nonlinearities g_{LO} for a range of noise characteristics is the class of *generalized Cauchy* densities. A generalized Cauchy density is defined in terms of three parameters σ^2, $k > 0$ and $v > 0$ by

$$f_{k,v}(x) = \frac{B(k,v)}{\left\{1 + \frac{1}{v}\left[\frac{|x|}{A(k)}\right]^k\right\}^{v+1/k}} \tag{3-9}$$

where

$$B(k,v) = \frac{k\, v^{-1/k}\, \Gamma(v+1/k)}{2A(k)\, \Gamma(v)\, \Gamma(1/k)} \tag{3-10}$$

and where $A(k)$ is defined by (3-2). The density function $f_{k,v}(x)$ has an algebraic rather than an exponential tail behavior. We see from (3-9) that the tails of the density function decay in inverse proportion to $|x|^{vk+1}$ for large $|x|$. For $k = 2$ and $v = 1/2$

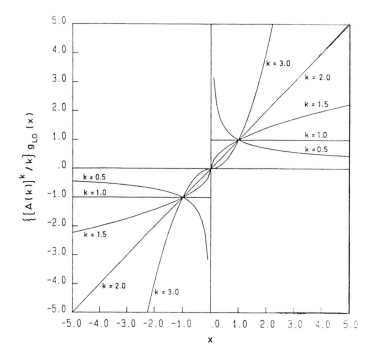

Figure 3.5 Normalized LO Nonlinearities for Generalized Gaussian PDF's

the density function becomes the Cauchy density,

$$f_{k,v}(x) = \frac{1}{\pi\sigma} \frac{1}{1 + (x/\sigma)^2} \qquad (3\text{-}11)$$

showing that σ^2 is in general a scale parameter but *not* the variance.

The Cauchy density itself does not have a finite variance. In spite of this non-physical property, it has been considered to be useful in modeling impulsive noise. One justification for this is that if a detector has acceptable performance in such noise, then it will most likely have acceptable performance in actual impulsive noise. In addition to being a generalization of the Cauchy noise density, the generalized Cauchy density of (3-9) includes as a special case a model for impulsive noise proposed by Mertz [1961]. From the amplitude density function of the impulsive noise proposed by Mertz, an assumption of symmetry of the noise density function leads to the generalized Cauchy density with $k = 1$. Furthermore, a model for impulsive noise proposed by Hall [1966]

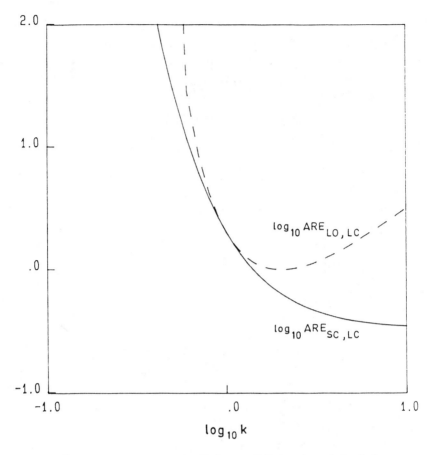

Figure 3.6 Asymptotic Relative Efficiencies of the LO Detectors and of the SC Detector, Relative to the LC Detector, for Generalized Gaussian Noise

is obtained with $k = 2$. When $2v$ is an integer this gives a scaled *Student-t* density.

It can be shown that the variance of the generalized Cauchy density, if it exists, is $\sigma^2 v^{2/k} \Gamma(v - 2/k)/\Gamma(v)$. This is finite for $vk > 2$, which is consistent with our observation that the tails of the density function decay in inverse proportion to $|x|^{vk+1}$. Figure 3.7 shows the shapes of the generalized Cauchy densities for different values of k when $v = 4$ and $\sigma^2 = 1$. For a fixed value of σ^2 and k, the limiting case $v \to \infty$ gives the generalized Gaussian density of (3-1). This can be inferred directly from the definition $e^a = \lim_{z \to \infty} (1 + a/z)^z$. Thus we also get the Gaussian density as a limiting case for $k = 2$ and $v \to \infty$. The generalized Cauchy density function is therefore capable of modeling a wide range of noise density types.

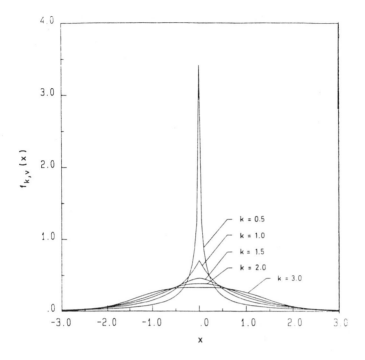

Figure 3.7 Generalized Cauchy Probability Density Functions ($\sigma^2 = 1$, $\upsilon = 4$)

The locally optimum nonlinearity g_{LO} for generalized Cauchy noise is easily found to be

$$g_{LO}(x) = \frac{\upsilon k + 1}{\upsilon [A(k)]^k + |x|^k} |x|^{k-1} \text{sgn}(x) \qquad (3\text{-}12)$$

which should be compared with g_{LO} for the generalized Gaussian densities of (3-4). The nonlinearity of (3-12) does approach, pointwise, that of (3-4) as $\upsilon \to \infty$. On the other hand, for any fixed set of values of the parameters υ, k, and σ^2 the characteristic g_{LO} of (3-12) always approaches zero for $|x| \to \infty$. This behavior is illustrated in Figure 3.8, which shows the normalized versions $\{\upsilon [A(k)]^k / (\upsilon k + 1)\} g_{LO}$ of g_{LO} for the generalized Cauchy densities for different values of the parameter k with $\sigma^2 = 1$ and $\upsilon = 10$. As we have remarked, for $\upsilon = 1/2$ and $k = 2$ we get the Cauchy noise density of (3-11), for which g_{LO} is

$$g_{LO}(x) = \frac{2x}{\sigma^2 + x^2} \qquad (3\text{-}13)$$

This function is plotted in Figure 3.9 for $\sigma^2 = 1$. One interesting conclusion we can draw from these plots is that there are several combinations of the parameters k and v which lead to LO nonlinearities of the type shown in Figure 3.3 for ELF noise; they are approximately linearly increasing near $x = 0$, and then decay to zero for large magnitudes of x.

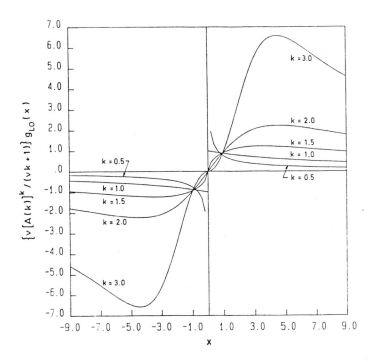

Figure 3.8 Normalized LO Nonlinearities for Generalized Cauchy PDF's ($\sigma^2 = 1$, $v = 10$)

The Fisher information $I(f_{k,v})$ for generalized Cauchy noise can be shown to be

$$I(f_{k,v}) = \frac{(vk + 1)^2 \, \Gamma(3/k) \, \Gamma(v + 1/k) \, \Gamma(v + 2/k) \, \Gamma(2 - 1/k)}{\sigma^2 v^{2/k} \, \Gamma^2(1/k) \, \Gamma(v) \, \Gamma(2 + v + 1/k)} \quad (3\text{-}14)$$

a finite quantity for $k > 0.5$. Note that σ^2 is not the variance of $f_{k,v}$ of (3-9); the variance is finite for $vk > 2$. From this we may calculate $ARE_{LO,LC}$, the ARE of the LO detector for $f_{k,v}$ relative to the LC detector, when $f_{k,v}$ is the noise density function. The result is shown in Figure 3.10, where $ARE_{LO,LC}$ is plotted as a function of k for different values of the parameter v. Once again

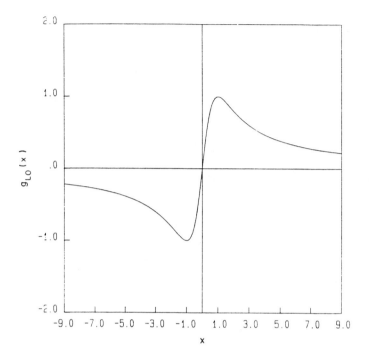

Figure 3.9 LO Nonlinearity for Cauchy Noise PDF ($\sigma^2 = 1$)

we find that substantial improvements in asymptotic performance can be obtained by use of LO or AO detectors if the noise density function is known to be a particular non-Gaussian density. The use of the simple hard-limiting SC detector can also be shown to produce a significant improvement over the LC detector for a range of values of the parameters k and v.

Before we go on to consider mixture noise densities, we should mention that one simple model which leads to some of the noise densities we have discussed above is the random-power-level Gaussian model. Here we model the noise samples W_i as products of standard Gaussian (zero-mean, unit-variance) components and random amplitude factors with various density functions. This is, in fact, the basis for the Hall model. Such noise models have also been used by Spooner [1968], Hatsell and Nolte [1971] and Adams and Nolte [1975]. The results we have described in this section were originally developed by Miller and Thomas [1972], who also considered the finite-support generalized beta densities which we have not included here. Some simple asymmetrical noise pdf models related to the Gaussian pdf are given in [Kanefsy and Thomas, 1965]. Huang and Thomas [1983] have also specifically

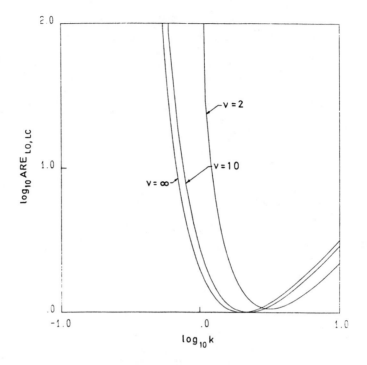

Figure 3.10 Asymptotic Relative Efficiencies of the LO Detectors, Relative to the LC Detector, for Generalized Cauchy Noise

considered nearly Gaussian skewed distributions and have compared the SC and LC detector performances under this condition.

3.3 Mixture Noise and Middleton Class A Noise

Suppose we want to model the behavior of a noise density function which is approximately Gaussian near the origin but has tails which decay at a lower rate than do the Gaussian density tails. An analytically simple model is provided by the mixture density function

$$f(x) = (1 - \epsilon)\eta(x) + \epsilon h(x) \tag{3-15}$$

where ϵ is some small positive constant, η is a Gaussian density function, and h is some other density function with heavier tails. Clearly, f defined by (3-15) is a valid density function as long as ϵ lies in the interval [0,1]. For small enough values of ϵ the behavior of f near the origin is dominated by that of η, assuming that h is a bounded function. For large values of $|x|$, however, h dominates the behavior of f since its tails decay at a slower rate than do those of η.

Mixture densities of the form of (3-15) have been used by many investigators to model heavy-tailed non-Gaussian noise density functions. In robustness studies they have been used to model classes of allowable noise density functions which are in the neighborhood of a nominal Gaussian density function η. The mixture model has also been found to provide a good fit, in many cases, to empirical noise data. As an example, we point to the work by Trunk [1976] in which the mixture model is used to fit the empirically determined probability distribution function of the envelope of radar clutter returns. In this case η in (3-15) is a Rayleigh rather than a Gaussian density. In particular, Trunk used what we call the *Gaussian-mixture* model (more accurately, Rayleigh mixture), with h in (3-15) also a Gaussian (or Rayleigh) density with a variance larger than that of η.

The mixture noise density model of (3-15) can also be given a justification as an appropriate model for impulse noise, considered to be a train of random-amplitude and randomly occurring narrow pulses in a background of Gaussian noise. Let us express such an impulsive component of a noise waveform as

$$I(t) = \sum_{k=-\infty}^{\infty} A_k p(t - t_k) \qquad (3\text{-}16)$$

Here the amplitudes A_k can be taken to be i.i.d. amplitudes and the t_k have commonly been assumed to be generated by a Poisson point process. The pulse shape p is determined by the receiver filter response. The classical analysis of Rice [1944] leads to a general result for the first-order probability density function of $I(t)$. Let v be the rate parameter of the Poisson point process and let T_p be the width of the pulse p. Then for $vT_p \ll 1$, one can derive the following approximation for the density function f_I of samples of $I(t)$ [Richter and Smits, 1971]:

$$f_I(x) = (1 - vT_p)\delta(x) + vT_p h_I(x) \qquad (3\text{-}17)$$

Here h_I is a density function which depends on the pulse shape p and on the density function of the A_k. The quantity $1 - vT_p$ may be viewed as the probability that no impulse noise is present at the sampling time; it comes from the Poisson assumption about the occurrence times t_k. The result of (3-17) is reasonable for the low-density case $vT_p \ll 1$, for which there are gaps between successive noise pulses in the impulsive component $I(t)$. Upon adding an independent Gaussian background noise process to $I(t)$ the first-order density function of the total noise process becomes a convolution of f_I with η, resulting in the noise density f of (3-15) in which ϵ is now vT_p and h is the convolution of h_I and η.

Middleton [1977, 1979a and 1979b, 1983] has described a canonical model based on a representation of the impulsive component of noise similar to, but more general than, that given by (3-16). For what is called his class A model the pulse widths are wide relative to the receiver filter impulse response so that the receiver filter passes the impulsive components, and physical mechanisms generating the interference can be used to interpret the parameters of the model for the noise. In terms of his two basic parameters (one related to vT_p above, the other being the ratio of noise powers in the impulsive and Gaussian components) Middleton has obtained an expansion of the noise density function f as an infinite weighted sum of Gaussian densities with decreasing weights for Gaussian densities with increasing variances. Specifically, it has been shown that the univariate probability density function of the normalized, unit-variance noise which has a Gaussian component and an independent additive interference component arising from a Poisson mechanism may be approximated as

$$f(x) = \sum_{m=0}^{\infty} \frac{e^{-A} A^m}{m!} \frac{1}{\sqrt{2\pi\sigma_m^2}} e^{-x^2/(2\sigma_m^2)} \qquad (3\text{-}18)$$

Here the parameter A, called the impulsive index, is like the product vT_p, being the product of an average rate of interfering waveform (pulse) generation and the waveform's mean duration. Thus a small value of A implies highly impulsive interference. The variances σ_m^2 are also related to physical parameters, being given as

$$\sigma_m^2 = \frac{(m/A) + \Gamma'}{1 + \Gamma'} \qquad (3\text{-}19)$$

where Γ' is the ratio of power in the Gaussian component of the noise to the power in the Poisson-mechanism interference. The major appeal of this model is that its parameters have a direct physical interpretation; the class A model has been found to provide very good fits to a variety of noise and interference measurements [Spaulding and Middleton, 1977a, Middleton, 1979a]. The Middleton Class A noise model is actually a model for *narrowband* noise, and it is often the *envelope* of such noise that is to be described statistically. The envelope probability density function for this noise is easily obtained from (3-18) as an infinite mixture of Rayleigh noise densities.

By limiting the sum in (3-18) to the first M terms only, and dividing by the sum of the first M coefficients of the Gaussian densities to maintain normalization, approximations of Middleton's class A model are obtained. It is of significance here that for several cases of empirically fitted noise densities a rather small value for the integer M is found to be sufficient to give excellent approximations to both the noise probability density functions and to the corresponding locally optimum nonlinearities. Such a numerical study has been reported recently by Vastola [1984]. From this observation one concludes that with $M = 2$ terms in (3-18), which does give very good approximations in the instances which have been considered, we end up with the mixture model of (3-15) in which $\epsilon = A/(1 + A)$. Assuming that η and h in (3-15) have zero means and respective variances σ_η^2 and σ_h^2, this also gives us that $\sigma_h^2/\sigma_\eta^2 = (1 + A\Gamma')/A\Gamma'$. In addition, of course, we find that h is also Gaussian here. Thus we find that the two-term mixture model of (3-15) has an additional justification as a special case arising from the class A model. In this particular interpretation the model parameters ϵ and σ_h^2/σ_η^2 can also be given a direct physical interpretation. For typical values of A and Γ, in the ranges (0.01, 0.5) and (0.0001, 0.1), respectively, we find that ϵ is in the range (0.01, 0.33) and σ_h^2/σ_η^2 is in the range (20, 10000). This indicates that the variance of the contaminating h in the mixture model may be much larger than that of the Gaussian component in real situations. We also remark here that Gaussian mixture models in general, of the type described by (3-18), may be thought of as special cases of the type of model mentioned at the end of the preceding section; this is the model of noise with a conditional density function which is Gaussian, conditioned on the variance, which is a random parameter.

Returning to the mixture model of (3-15) in which both η and h have zero means, we find that the variance σ^2 of the mixture noise density is

$$\sigma^2 = (1 - \epsilon)\sigma_\eta^2 + \epsilon\sigma_h^2 \tag{3-20}$$

and the LO nonlinearity is

$$g_{LO}(x) = \frac{-(1-\epsilon)\eta'(x) - \epsilon h'(x)}{(1-\epsilon)\eta(x) + \epsilon h(x)} \tag{3-21}$$

Now we can express this in two alternative ways, using $\eta'(x) = -(x/\sigma_\eta^2)\eta(x)$:

$$g_{LO}(x) = \frac{x/\sigma_\eta^2 - [\epsilon/(1-\epsilon)]h'(x)/\eta(x)}{1 + [\epsilon/(1-\epsilon)]h(x)/\eta(x)} \tag{3-22}$$

$$g_{LO}(x) = \frac{-h'(x)/h(x) + [(1-\epsilon)/\epsilon](x/\sigma_\eta^2)\eta(x)/h(x)}{1 + [(1-\epsilon)/\epsilon]\eta(x)/h(x)} \quad (3\text{-}23)$$

Assuming that h' as well as h is bounded, for $\epsilon \ll 1$ we find from (3-22) that for small values of $|x|$ the function g_{LO} is approximately linear in x. If $h'(0)$ is not zero, then for very small values of $|x|$ we will get a different behavior for g_{LO}, since $g_{LO}(0)$ will then have a value of approximately $-\epsilon h'(0)/\eta(0)$. Note that $h'(0)$ is strictly speaking $h'(0+)$ or $h'(0-)$. For large values of $|x|$ the behavior of g_{LO} is revealed more easily by (3-23). If the tails of h decay at a lower rate than do those of η, in such a way that $x\eta(x)$ goes to zero faster than do $h(x)$ and $h'(x)$ for $|x| \to \infty$, the second terms in the numerator and denominator in (3-23) become negligible for large values of $|x|$. Thus the limiting behavior of g_{LO} for large $|x|$ will generally be that of the LO characteristic for noise density function h, for heavy-tailed h.

Let h be the double-exponential noise density function of (2-15), in which the parameter a is related to the variance σ_h^2 by $\sigma_h^2 = 2/a^2$. Figure 3.11 shows the characteristic g_{LO} for the Gaussian and double-exponential mixture density with $\epsilon = 0.05$ and $\sigma_h^2/\sigma_\eta^2 = 10$. The general characteristics of g_{LO} are as we would expect them to be. Note once again that the shape of g_{LO} here is similar to that determined for ELF noise, as given in Figure 3.3.

These results once again indicate that the simple sign correlator detector should perform well in impulsive noise modeled to have the mixture density function, as compared to the linear correlator detector. The ARE of the SC relative to the LC detector for the mixture noise density is given by

$$\begin{aligned} ARE_{SC,LC} &= 4\sigma^2 f^2(0) \\ &= 4[(1-\epsilon)\sigma_\eta^2 + \epsilon\sigma_h^2]\left[\frac{1-\epsilon}{\sqrt{2\pi\sigma_\eta^2}} + \epsilon h(0)\right]^2 \\ &= 4[1 - \epsilon + \epsilon\sigma_h^2/\sigma_\eta^2]\left[\frac{1-\epsilon}{\sqrt{2\pi}} + \sigma_\eta \epsilon h(0)\right]^2 \quad (3\text{-}24) \end{aligned}$$

Note that the factor containing $h(0)$ above has a minimum value of $(1-\epsilon)^2/2\pi$, which is larger than zero for $\epsilon < 1$. It follows then that for $\epsilon > 0$ as well, $ARE_{SC,LC}$ can be made arbitrarily large by making σ_h^2/σ_η^2 sufficiently large. This is true specifically when h is the double-exponential density function. For this contamination noise density function h the $ARE_{SC,LC} = 2$ when $\epsilon = 1$, in which case the SC detector is the LO detector. It is thus interesting that for any ϵ strictly between zero and unity the $ARE_{SC,LC}$ can

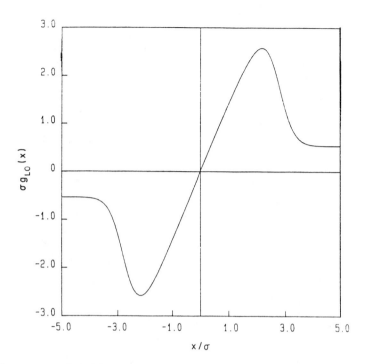

Figure 3.11 LO Nonlinearity for Gaussian and Double-Exponential Noise PDF Mixture ($\epsilon = 0.05$, $\sigma_h^2/\sigma_n^2 = 10$)

become arbitrarily large for such an h. The explanation is that the efficacy of the sign correlator detector is $4f^2(0)$, which for the mixture density always contains a fixed contribution due to η for any σ_h^2. The quantity $4h^2(0)$ by itself decreases to zero for $\sigma_h^2 \to \infty$. The mixture density function f is *not* simply *scaled* by σ_h^2 for $0 < \epsilon < 1$. Note that the LC detector, on the other hand, has efficacy $1/\sigma^2$, which converges to zero for $\sigma_h^2 \to \infty$.

The factor $[1 - \epsilon + \epsilon \sigma_h^2/\sigma_\eta^2]$ in (3-24) also has a minimum value of $1 - \epsilon$ which is positive for $\epsilon < 1$. Now $\sigma_\eta \epsilon h(0) = (\sigma_\eta/\sigma_h)\epsilon h(0)\sigma_h$, and $h(0)\sigma_h$ has a constant value which we shall assume is not zero. Thus we find that for $0 < \epsilon < 1$ the $ARE_{SC,LC}$ also approaches ∞ for $\sigma_h^2/\sigma_\eta^2 \to 0$. For any fixed σ_η^2, the efficacy $1/\sigma^2$ of the LC detector approaches the value $[(1-\epsilon)\sigma_\eta^2]^{-1}$ for $\sigma_h^2 \to 0$. The efficacy $4f^2(0)$ of the SC detector now contains a contribution due to $\epsilon h(0)$ which, on the other hand, approaches ∞ as σ_h converges to zero, since $h(0)\sigma_h$ is some positive constant. When $\epsilon = 0$ the $ARE_{SC,LC}$ becomes $2/\pi$, its value for Gaussian noise.

We thus see that the SC detector can far outperform the LC detector in highly impulsive noise modeled by the mixture density, for which $\sigma_\kappa^2/\sigma_\eta^2$ is high, and also in the other extreme case of the mixture model placing a relatively high probability near the origin (Gaussian noise with "quiet" periods). For some numerical evaluations of the $ARE_{SC,LC}$ we refer the reader to the work of Miller and Thomas [1976].

One general conclusion we can make from the specific cases of non-Gaussian noise we have considered here is that with appropriate nonlinear processing of the inputs it is possible to get a substantial advantage over detectors employing the LC test statistic. In all our work we have assumed the availability of independent data components X_i at the detector input. This in turn is obtained under the assumption that the noise is almost "white" relative to the low-pass signal bandwidth. It is important to keep in mind that it is *not* permissible first to low-pass filter (or narrowband filter) the input process if it contains impulsive non-Gaussian noise. This is not a problem for Gaussian noise, since the very operation of linear-correlation amounts to use of a low-pass filter in any case. For non-Gaussian white noise it is the nonlinear limiting type of operation performed by g_{LO} which accounts for the major improvement in performance, through the de-emphasis of large noise values which this implies. The resulting nonlinearly processed data is then effectively low-pass filtered when the GC test statistic is formed. Thus it is permissible to insert an ideal low-pass filter passing the signal frequencies *after* the zero-memory nonlinearity g_{LO}. But interchanging the order of these functions leads to a different system; the low-pass filter spreads out noise impulses in time, thus very significantly reducing the advantage obtainable with further nonlinear processing. It is for this reason that in practical systems such as that mentioned for ELF noise, pre-whitening or equalization filters are used before hard-limiting of the data to reverse the effect of any low-pass filtering which may have taken place.

One useful indication the results of this and the previous section give us is that detector nonlinearities g which are approximately linear near the origin but which become saturated and tend to limit the influence of large-magnitude observations should give good overall performance for both Gaussian and heavier-tailed impulsive noise density functions. We find, indeed, that such a nonlinearity can give us a *minimax robust* detection scheme for noise densities modeled by the mixture model (Kassam and Poor [1985]). Other noise models and detector performance studies related to our discussion here may be found in the works of Kurz [1962], Rappaport and Kurz [1966], Klose and Kurz [1969], Chelyshev [1973], Freeman [1981] and Aazhang and Poor [1985, 1986]. Many experimental studies of noise processes arising in various applications have been reported; we refer here to the work of Milne and Ganton [1964] as one example, indicating the highly impulsive nature of underwater noise under certain conditions.

3.4 Adaptive Detection

We have so far always assumed, in obtaining the LO detectors and evaluating their performances, that the noise density function f is completely specified. In this book, in fact, our considerations are primarily confined to such situations and to *fixed* or *non-adaptive* detectors. At least in one special case of incompletely specified noise density an adaptive scheme is quite useful. This is the case where f is known except for its variance σ^2, the noise power. In this case the LO nonlinearity g_{LO} has a known shape, but its input-scaling factor is unknown. More generally, if f is not known, it is unlikely that g_{LO} can be effectively estimated and adaptively updated. But a more reasonable scheme may be possible, in which the test statistic is chosen to be of some simple form with only a few variable parameters. These may then be adaptively set to optimize performance under different noise conditions. We shall focus briefly on this latter topic here, but let us discuss the former first.

Unknown Noise Variance

Let the noise density function be defined as

$$f(x) = \frac{1}{\sigma}\overline{f}(x/\sigma) \tag{3-25}$$

where $\overline{f}$ is a known unit-variance density and σ^2 is an unknown parameter, the variance of f. The LO characteristic is that given by

$$g_{LO}(x) = -\frac{1}{\sigma}\frac{\overline{f}'(x/\sigma)}{\overline{f}(x/\sigma)}$$

$$= \frac{1}{\sigma}\overline{g}_{LO}(x/\sigma) \tag{3-26}$$

Note that $\overline{g}_{LO} = -\overline{f}'/\overline{f}$ is now a known function, but g_{LO} is an amplitude as well as input scaled version of $\overline{g}_{LO}$. The amplitude scaling is not essential, and we may write down the LO test statistic as

$$\lambda_{LO}(\mathbf{X}) = \sum_{i=1}^{n} s_i \overline{g}_{LO}(X_i/\sigma) \tag{3-27}$$

Under the noise-only null hypothesis H_1 of (2-13) the density function of X_i/σ is simply $\bar{f}$, so that the null-hypothesis distribution of $\lambda_{LO}(\mathbf{X})$ is exactly known and the detector threshold can be obtained for any false-alarm probability specification. To implement $\lambda_{LO}(\mathbf{X})$ it is necessary to have a good estimate of σ. If the noise power level is non-stationary, periodic updating of the variance is also necessary. Adaptive detection schemes which use input scaling with estimates of σ are quite commonly used in practice, and are commonly refered to as AGC (automatic gain control) schemes. Of course it is important that good estimates of σ be available; otherwise, significant departure from expected performance may result. A related adaptation strategy has been considered by Lu and Eisenstein [1984].

Unknown Noise Density

A useful adaptive detector structure is obtained by allowing only a few variable parameters in the definition of the detector test statistic, which may then be set adaptively to match the noise conditions. One particularly simple structure is suggested by our results of the previous sections, where we saw that the sign correlator detector generally performs quite well for heavy-tailed noise density functions which are typical for impulsive noise. On the other hand, for Gaussian noise the LC detector is optimum and the SC detector performs rather poorly. Now consider the test statistic

$$T_{GC}(\mathbf{X}) = \sum_{i=1}^{n} s_i [\gamma X_i + (1-\gamma) \operatorname{sgn}(X_i)]$$
$$= \gamma \sum_{i=1}^{n} s_i X_i + (1-\gamma) \sum_{i=1}^{n} s_i \operatorname{sgn}(X_i) \quad (3\text{-}28)$$

in which γ is a free parameter. The choice $\gamma = 1$ makes this an LC test statistic, and $\gamma = 0$ makes it the SC test statistic. Therefore, if an adaptive scheme can be formulated which allows γ to be picked optimally for different noise density conditions, one can expect the resulting detector to have very good performance over a range of noise conditions.

This type of test statistic was suggested by Modestino [1977], from whose results it follows that the normalized efficacy $\tilde{\xi}_\gamma$ of $T_{GC}(\mathbf{X})$ of (3-28) for our detection problem is

$$\tilde{\xi}_\gamma = \frac{[\gamma + 2(1-\gamma)f(0)]^2}{\gamma^2 \sigma^2 + 2\gamma(1-\gamma)p_0 + (1-\gamma)^2} \quad (3\text{-}29)$$

where

$$p_0 = \int_{-\infty}^{\infty} |x| f(x)\, dx \quad (3\text{-}30)$$

This result for $\tilde{\xi}_\gamma$ is quite simple to derive. Modestino [1977] shows that $\tilde{\xi}_\gamma$ is maximized for

$$\gamma = \frac{1 - 2p_0 f(0)}{(1 - p_0) + 2f(0)(\sigma^2 - p_0)} \qquad (3\text{-}31)$$

and the maximum value of $\tilde{\xi}_\gamma$ using this value for γ is

$$\tilde{\xi}_{\max} = \frac{1 - 4p_0 f(0) + 4\sigma^2 f^2(0)}{\sigma^2 - p_0^2} \qquad (3\text{-}32)$$

For the generalized Gaussian noise densities of (3-1), evaluation of $\tilde{\xi}_{\max}$ shows that the test statistic of (3-28) with optimum choice of γ has very good asymptotic performance relative to the AO detectors for $1 \leq k \leq 2$, where k is the rate of exponential decay parameter for the generalized Gaussian densities. In fact, the ARE of the optimum detector based on $T_{GC}(\mathbf{X})$ of (3-28) relative to the AO detector is always larger than 0.95 for this range of values.

One possibility in devising an adaptive scheme using $T_{GC}(\mathbf{X})$ of (3-28) is to estimate the parameters p_0, $f(0)$ and σ^2; this requires noise-only training observations. Another possibility is to use a stochastic approximation technique operating on signal-plus-noise observations, which attempts to maximize the output SNR for $T_{GC}(\mathbf{X})$. This approach has been discussed by Modestino [1977]. A detector structure related to the one we have discussed above has been considered in [Czarnecki and Thomas, 1983].

A more general approach is to start with a specified set of nonlinearities $\{g_1(x), g_2(x),...,g_M(x)\}$, and to consider as possible detector test statistics all linear combinations of these M functions. An adaptive detector may then be sought which uses coefficients for its linear combination which are optimum, say in the sense of maximizing the detector efficacy, for the prevailing noise probability density function. A useful simplification is obtained if the functions $g_j(x)$, $j = 1,2,...,M$, form an *orthonormal* set for the allowable noise probability density functions. This means that for allowable f we have

$$\int_{-\infty}^{\infty} g_j(x) g_k(x) f(x)\, dx = \delta_{jk} \qquad (3\text{-}33)$$

where δ_{jk} is the Kronecker delta function. Under this constraint the detector efficacy has a simple expression and the conditions on the linear combination coefficients to maximize efficacy can be found easily (Problem 2.8, Chapter 2). These conditions can also form the basis for an adaptive detection scheme maximizing performance in different noise environments. A special case is obtained when the orthonormal functions $g_j(x)$ are defined to be non-zero over corresponding sets P_j, $j = 1,2,...,M$, which are *disjoint*. An important example of this is quantization, in which the collection of M functions $\{g_j,\ j = 1,2,...M\}$ together with the set of linear combination weighting coefficients define a quantizer characteristic. We shall discuss these ideas further at the end of the next chapter.

While it is possible, then, to consider some simplified schemes for adaptive detection in unknown and nonstationary noise environments, the implementation of an efficient adaptive scheme necessarily adds, often significantly, to the complexity of the detector. Furthermore, there are considerations such as of sample sizes required to achieve convergence which have to be taken into account in designing a useful scheme of this type.

PROBLEMS

Problem 3.1

An alternative parameterization of the generalized Gaussian pdf's of (3-1) is in terms of k and $a = A(k)$, giving

$$f_k(x) = \frac{k}{2a\,\Gamma(1/k)}\, e^{-[|x|/a]^k}$$

Note that $A(k)$ was defined in (3-2) in terms of σ^2, the variance.

(a) Show that for fixed value of a the pdf f_k converges for $k \to \infty$ to the *uniform* pdf on $[-a, a]$.

(b) Find the limit to which $ARE_{SC,LC}$ of (3-8) converges, as $k \to \infty$. Verify that this agrees with the result of a direct calculation of $ARE_{SC,LC}$ for the uniform pdf.

Problem 3.2

Let V be a Gaussian random variable with mean zero and variance unity. Let Z be an independent Rayleigh random variable with pdf

$$f_Z(z) = \frac{z}{a^2}\, e^{-z^2/2a^2},\ z \geq 0$$

Find the pdf of $X = ZV$. (This is a model for a conditionally Gaussian noise sample with a random power level.)

Problem 3.3

The symmetric, unimodal *logistic* pdf is defined as
$$f(x) = \frac{1}{a} \frac{e^{-x/a}}{[1 + e^{-x/a}]^2}, \quad -\infty < x < \infty$$

(a) Prove that its variance is $a^2 \pi^2/3$. (Use a series expansion.) Plot on the same axes the zero-mean unit-variance Gaussian, double-exponential, and logistic pdf's. (The logistic pdf is useful as a non-Gaussian pdf having exponential tail behavior and smooth behavior at the origin.)

(b) Show that g_{LO} for this pdf is given by
$$g_{LO}(x) = \frac{1}{a} [2 F(x) - 1]$$

where F is the cumulative distribution function corresponding to the pdf f above. Sketch $g_{LO}(x)$.

(c) Find $ARE_{LC,LO}$ and $ARE_{SC,LO}$ in testing H_1 versus K_1 of (2-13) and (2-14) when f is the logistic pdf.

Problem 3.4

Obtain $ARE_{SC,LC}$ for generalized Cauchy noise in (2-13) and (2-14), and sketch the result as a function of k for a small, an intermediate, and a large value of v.

Problem 3.5

In the mixture model of (3-15) let η and h both be zero-mean Gaussian pdf's, with η having unit variance. For $\sigma_h^2 = 100$ and $\epsilon = 0.05$ plot the LO nonlinearity using log scales for both axes.

Problem 3.6

Let f be the mixture noise pdf of (3-15) in which η is the zero-mean, unit-variance Gaussian pdf and h is a bounded symmetric pdf. Consider the GC detector based on the soft-limiter function defined in Problem 2.5 of Chapter 2; we will denote this soft-limiter function as l_c.

(a) For fixed c, find the mixture proportion ϵ and the pdf h (in terms of η and c) for which the resulting mixture pdf $f = f^*$ has l_c as the LO nonlinearity g_{LO}.

(b) Consider the class of mixture pdf's obtained with different bounded symmetric h in the mixture model with η fixed as above and with ϵ fixed in terms of c and η as above. Show that the efficacy of the GC detector based on l_c, for H_1 versus K_1, is a *minimum* for the pdf f^* in this class. (This is the essence of the min-max robustness property of the soft limiter.)

Chapter 4

OPTIMUM DATA QUANTIZATION IN KNOWN-SIGNAL DETECTION

4.1 Introduction

In this section we will continue to consider the known signal in additive noise detection problem, but we will now be interested in detection based on quantized data. Instead of using the observations X_i, $i = 1,2,...,n$, to decide between $\theta = 0$ and $\theta > 0$ in the model of (2-1), we will now consider detectors which use quantized versions $q(X_i)$, where q is some piecewise-constant function.

The need to use quantized data can arise in many situations. If remotely acquired data are to be transmitted to a central processing facility, for example, data rate reduction may be highly desirable or necessary. Similarly, storage of data for off-line processing may also require quantization into a relatively small number of levels, especially if one is dealing with a large volume of data. The use of quantized data also turns out to produce more *robust* detectors which are not very sensitive in their performance to deviations of the actual noise density functions from those assumed in their design. Finally, the use of quantized data can allow simple adaptive schemes to be implemented to allow the detector to retain good performance even when the noise characteristics can vary considerably over time.

The problem of optimum approximation of analog data using a multilevel or digital representation with a finite number of bits per analog sample is an old problem with well-known solutions. Classic among these is the solution of Max [1960], which gives the optimum quantizer minimizing the *mean-squared distortion* between the analog sample and its quantized version. If Y is an analog quantity whose observed value is governed by some probability density function f_Y, Max's solution allows us to obtain the optimum M-interval quantizer q which minimizes the mean-squared error $e(q;f_Y) = E\{[Y - q(Y)]^2\}$. Note that this can be written as

$$e(q;f_Y) = \int_{-\infty}^{\infty} [y - q(y)]^2 f_Y(y)\, dy$$

$$= \sum_{j=1}^{M} \int_{t_{j-1}}^{t_j} (y - l_j)^2 f_Y(y)\, dy \qquad (4\text{-}1)$$

where l_j is the output level of the quantizer q when the input is

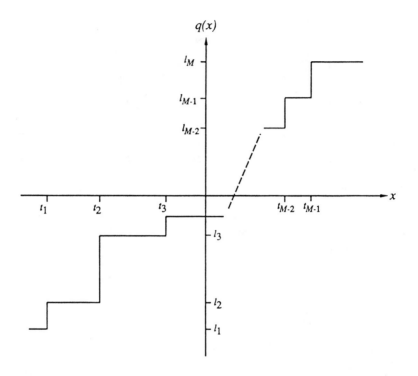

Figure 4.1 General M-Interval Quantizer Characteristic

in the interval $(t_{j-1}, t_j]$. For M-interval quantization we have $t_M = \infty$; we always set $t_0 = -\infty$. The general quantizer characteristic q is shown in Figure 4.1. The necessary conditions on the levels l_j and breakpoints t_j for q to minimize $e(q;f_Y)$ are obtained by differentiation, and are

$$t_j = \frac{l_j + l_{j+1}}{2}, \quad j = 1,2,...,M-1 \qquad (4\text{-}2)$$

and

$$l_j = \frac{\int_{t_{j-1}}^{t_j} y f_Y(y)\, dy}{F_Y(t_j) - F_Y(t_{j-1})}, \quad j = 1,2,...,M \qquad (4\text{-}3)$$

where F_Y is the distribution function of Y. We see that the optimum j-th level is the mean value of Y conditioned on it being in the j-th input interval, and the optimum j-th breakpoint is the average value of the j-th and $(j+1)$-th levels. Equations (4-2) and (4-3) have to be solved simultaneously, which can be accomplished using iterative numerical techniques, to obtain a minimum mean-squared-error quantizer (assuming f_Y is sufficiently regular so that the necessary conditions are also sufficient for local minimization). We shall call such a quantizer a *minimum distortion* quantizer.

In the detection context where the X_i arise from the model of (2-1), it is not at all clear how we can use a distortion criterion such as the mean-squared error as a basis for optimum quantizer design. To begin with, the probability density function of the X_i depends on whether the null or the alternative hypothesis is true. Since our objective is to use the quantized data in a signal detector, the quantizer should be chosen to extract the maximum amount of information most relevant in the detection problem. It is therefore more appropriate to use as a quantization performance criterion some measure of the detection performance obtained with the quantized data, and to design optimum schemes based on such a criterion.

For this purpose we will assume as our basic structure the system of Figure 4.2, in which instantaneously quantized data $q(X_i)$ are used to form a detection test statistic. The quantizer q is determined by its breakpoints t_j and output levels l_j, as shown in Figure 4.1. In an actual system the quantizer may, for example, be located at the point of data acquisition and the detector may be physically removed from it. The quantized data may then be transmitted over a communication link as a sequence of binary digits, and the output levels l_j would be reconstructed at the detector. The communication link may contain or be completely replaced by digital storage. From this we see that in general the essential quantization operation is that of *data reduction*, which corresponds to assignment of the input X_i to one of the M intervals $(t_{j-1}, t_j]$ of the input partitioning determined by the quantizer breakpoints. The detector, presented with such reduced data, may implicitly reconstruct output levels l_j for the reduced observations and use them in computing a test statistic. This was the assumption made in starting with the structure of Figure 4.2.

4.2 Asymptotically Optimum Quantization

Let us begin by assuming that the breakpoints t_j, $j = 0,1,...,M$, are specified (with $t_0 = -\infty$ and $t_M = \infty$) for M-interval quantization. The reduced data due to this input partitioning of the X_i into M intervals may be represented as the M-component row vector $\mathbf{Z}_i$, defined so that its j-th component Z_{ij} is

$$Z_{ij} = \begin{cases} 1 & , \quad t_{j-1} < X_i \leq t_j \\ 0 & , \quad \text{otherwise} \end{cases} \quad j = 1,2,...,M \qquad (4\text{-}4)$$

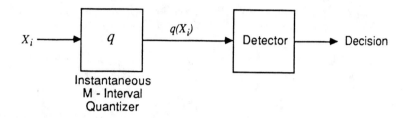

Figure 4.2 Detector Using Quantized Data

Thus each Z_i has a single non-zero entry of "1" in the position corresponding to the input interval in which X_i lies. Let Z be the $n \times M$ *matrix* of the n row vectors Z_i.

The joint probability density function of the components of Z can easily be seen to be

$$p_Z(z \mid \theta) = P\{Z_{ij} = z_{ij}, \begin{matrix} i = 1,2,\ldots,n \\ j = 1,2,\ldots,M \end{matrix}\}$$
$$= \prod_{i=1}^{n} \prod_{j=1}^{M} [p_{ij}(\theta)]^{z_{ij}} \qquad (4\text{-}5)$$

where

$$p_{ij}(\theta) = P\{t_{j-1} < X_i \leq t_j\}$$
$$= F(t_j - \theta s_i) - F(t_{j-1} - \theta s_i) \qquad (4\text{-}6)$$

This result is based on the assumption of (2-14), that the X_i are independent and identically distributed. Note that (4-5) is valid only for z such that only one element in each row has the value of unity, all others being zero; otherwise, $p_Z(z \mid \theta) = 0$.

From (4-5) and (4-6) we can obtain the joint probability density function of the reduced data matrix Z for $\theta = 0$ and, in general, for any arbitrary value $\theta = \theta_0 > 0$. To test $\theta = 0$ versus $\theta = \theta_0 > 0$ based on Z, the optimum detector can use the log-likelihood ratio

$$\ln L(Z) = \ln \prod_{i=1}^{n} \prod_{j=1}^{M} \left[\frac{p_{ij}(\theta_0)}{p_{ij}(0)} \right]^{Z_{ij}}$$
$$= \sum_{i=1}^{n} \sum_{j=1}^{M} Z_{ij} \ln \left[\frac{p_{ij}(\theta_0)}{p_{ij}(0)} \right] \qquad (4\text{-}7)$$

Notice that this test statistic is dependent on θ_0 in a non-trivial way, which therefore has to be known in order that $\ln L(\mathbf{Z})$ be obtained explicitly (for a given set of breakpoints t_j). We are thus faced with the same difficulty we encountered earlier in using the analog data X_i, in that a UMP quantization scheme cannot be found. As we did then, we may look for a *locally optimum* detector using the reduced data $\mathbf{Z}$. Again from the generalized Neyman-Pearson lemma that we have utilized in Chapter 2, using an argument similar to that giving the result (2-31) and hence (2-32), we find that the locally optimum detector D_{LO} based on the reduced data $\mathbf{Z}$ uses the test statistic

$$\lambda_{LO}(\mathbf{Z}) = \frac{\frac{d}{d\theta} p_\mathbf{Z}(\mathbf{Z}|\theta)\Big|_{\theta=0}}{p_\mathbf{Z}(\mathbf{Z}|\theta)\Big|_{\theta=0}}$$

$$= \frac{d}{d\theta} \ln p_\mathbf{Z}(\mathbf{Z}|\theta)\Big|_{\theta=0}$$

$$= \sum_{i=1}^{n} \sum_{j=1}^{M} Z_{ij} \frac{\frac{d}{d\theta} p_{ij}(\theta)\Big|_{\theta=0}}{p_{ij}(0)}$$

$$= -\sum_{i=1}^{n} \sum_{j=1}^{M} Z_{ij} \, s_i \, \frac{f(t_j) - f(t_{j-1})}{F(t_j) - F(t_{j-1})}$$

$$= \sum_{i=1}^{n} s_i \sum_{j=1}^{M} Z_{ij} \frac{f(t_{j-1}) - f(t_j)}{F(t_j) - F(t_{j-1})} \tag{4-8}$$

In view of the definition of the Z_{ij}, we can express $\lambda_{LO}(\mathbf{Z})$ in the form

$$\lambda_{LO}(\mathbf{Z}) = \sum_{i=1}^{n} s_i \, q_{\mathbf{t},LO}(X_i) \tag{4-9}$$

where $q_{\mathbf{t},LO}$ is the M-interval quantizer with the breakpoints vector $\mathbf{t} = (t_1, t_2, ..., t_{M-1})$ and the j-th output level l_j for input in $(t_{j-1}, t_j]$ being given by

$$l_j = \frac{f(t_{j-1}) - f(t_j)}{F(t_j) - F(t_{j-1})}, \quad j = 1, 2, \ldots, M \tag{4-10}$$

We see that $q_{t,LO}$ represents the locally optimum assignment of output levels in quantization with a given breakpoints vector **t**. Furthermore, note that the quantized-data test statistic of (4-9) is also a special case of a GC detector.

What we have accomplished so far is the establishment of the locally optimum output levels for a quantizer with specified breakpoints vector **t** operating on the observations X_i. To complete the specification of the LO quantization we need to determine the optimum choice for the breakpoints $t_1, t_2, \ldots, t_{M-1}$ for a given number M of input intervals. This would require a determination of the power function of the detector using $q_{t,LO}$ (for a particular false-alarm probability) whose slope at $\theta = 0$ would then have to be maximized with respect to the t_j. This becomes rather difficult to carry out in general for any noise density function f and sample size n. As an alternative, we can look for the *asymptotically optimum* values for the $M-1$ breakpoints for the sequence of hypothesis-testing problems $\{H_{1,n}$ versus $K_{1,n}, n = 1,2,\ldots\}$ with $\theta_n = \gamma/\sqrt{n}$, as in Section 2.3 of Chapter 2. We can obtain the AO breakpoints by maximizing the efficacy of the detectors based on $\lambda_{LO}(\mathbf{Z})$ of (4-9).

Let us apply the Pitman-Noether formula (1-41) to find the efficacy ξ_{QC} of any general *quantizer correlator* (QC) detector based on a QC test statistic of the form

$$T_{QC}(\mathbf{X}) = \sum_{i=1}^{n} s_i \, q(X_i) \tag{4-11}$$

The efficacy is found to be

$$\xi_{QC} = \lim_{n \to \infty} \frac{\left[\sum_{i=1}^{n} s_i \, \frac{d}{d\theta} E\{q(X_i) \mid K_1\} \Big|_{\theta=0} \right]^2}{n V\{q(X_i) \mid H_1\} \sum_{i=1}^{n} s_i^2} \tag{4-12}$$

which, since

$$\frac{d}{d\theta} E\{q(X_i) \mid K_1\} \Big|_{\theta=0} = -s_i \sum_{j=1}^{M} \int_{t_{j-1}}^{t_j} l_j f'(x) \, dx$$

$$= s_i \sum_{j=1}^{M} l_j [f(t_{j-1}) - f(t_j)] \tag{4-13}$$

becomes

$$\xi_{QC} = P_s^2 \frac{\left\{\sum_{j=1}^{M} l_j [f(t_{j-1}) - f(t_j)]\right\}^2}{\sum_{j=1}^{M} l_j^2 [F(t_j) - F(t_{j-1})] - \left\{\sum_{j=1}^{M} l_j [F(t_j) - F(t_{j-1})]\right\}^2}$$

$$= P_s^2 \frac{\left\{\sum_{j=1}^{M} (l_j - l_0)[f(t_{j-1}) - f(t_j)]\right\}^2}{\sum_{j=1}^{M} (l_j - l_0)^2 [F(t_j) - F(t_{j-1})]} \qquad (4\text{-}14)$$

Here

$$l_0 = E\{q(X_i) \mid H_1\}$$
$$= \sum_{j=1}^{M} l_j [F(t_j) - F(t_{j-1})] \qquad (4\text{-}15)$$

and P_s^2 is the average known-signal power defined in (2-44). This result could also have been derived using (2-77). From this we find that the efficacy $\xi_{QC,LO}$ of the QC detector using the LO output levels for a given breakpoints vector **t** is, from (4-10) and (4-14),

$$\xi_{QC,LO} = P_s^2 \sum_{j=1}^{M} \frac{[f(t_{j-1}) - f(t_j)]^2}{F(t_j) - F(t_{j-1})} \qquad (4\text{-}16)$$

Note that for the LO quantizer $q_{t,LO}$ using LO levels l_j of (4-10), the mean value l_0 of (4-15) is zero.

Before continuing let us observe that although the l_j of (4-10) were obtained as the LO levels for given **t**, they should also be the AO levels for a given **t** vector for detectors of the QC type using $T_{QC}(\mathbf{X})$ of (4-11) as test statistics. That this is indeed so can be verified by seeking the levels l_j which maximize the general efficacy expression ξ_{QC} of (4-14). Let us first note that ξ_{QC} is invariant to transformations of the levels from l_j to $al_j + b$, $j = 1,2,...,M$, for any $a \neq 0$ and b. Then it is quite easy to see (Problem 4.1) that with l_j defined in terms of **t** by (4-10), any set of levels of the form $al_j + b$, $j = 1,2,...,M$, with $a \neq 0$, maximizes ξ_{QC} for given **t**. This maximum value is $\xi_{QC,LO}$ of (4-16).

Continuing with maximization of $\xi_{QC,LO}$ with respect to the breakpoints, from the necessary conditions

$$\frac{\partial}{\partial t_j} \xi_{QC,LO} = 0, \; j = 1,2,...,M-1 \qquad (4\text{-}17)$$

for AO breakpoints maximizing $\xi_{QC,LO}$, we find that (Problem 4.2)

$$g_{LO}(t_j) = \frac{1}{2}\left\{\frac{f(t_{j-1}) - f(t_j)}{F(t_j) - F(t_{j-1})} + \frac{f(t_j) - f(t_{j+1})}{F(t_{j+1}) - F(t_j)}\right\},$$
$$j = 1,2,...,M-1 \qquad (4\text{-}18)$$

must be satisfied by the AO breakpoints. Here $g_{LO}(t_j)$ is $-f'(t_j)/f(t_j)$. Using (4-10), this last result becomes

$$g_{LO}(t_j) = \frac{l_j + l_{j+1}}{2}, \quad j = 1,2,...,M-1 \qquad (4\text{-}19)$$

The equations (4-19) above and (4-10) for the levels must be solved simultaneously for the output levels and breakpoints of an AO quantization of data. These two sets of equations provide a total of $2M - 1$ conditions which can be solved, generally numerically, for the M output levels and $M - 1$ breakpoints $t_1, t_2, ..., t_{M-1}$ for an AO quantizer.

A very interesting comparison can be made between these results and those on minimum distortion quantization summarized in the Introduction above. For this let us observe that the levels of (4-10) may be expressed as

$$l_j = \frac{\int_{t_{j-1}}^{t_j} g_{LO}(x) f(x)\, dx}{F(t_j) - F(t_{j-1})}, \quad j = 1,2,...,M \qquad (4\text{-}20)$$

which is the average value of g_{LO} on (t_{j-1}, t_j). Comparing (4-2) and (4-3) with (4-19) and (4-20), we see immediately that the quantizer which minimizes the mean-squared error between the quantizer characteristic and the *locally optimum nonlinearity* g_{LO} is also an asymptotically optimum quantizer. Notice that once (4-19) and (4-20) have been solved simultaneously for the AO quantizer parameters, the levels may be scaled and shifted by fixed scale and shift constants without changing the asymptotic optimality of the quantization; one would simply modify the detector threshold in the same way. The particular solution of the simultaneous equations (4-19) and (4-20) is the one giving also a quantizer which is the *best fit* to g_{LO} under noise-only conditions. More explicitly, we find that the mean-squared error $e(g_{LO}, q; f)$ between g_{LO} and q under H_1 is given by

$$e(g_{LO}, q; f) = E\{[g_{LO}(X_i) - q(X_i)]^2 \mid H_1\}$$

$$= \sum_{j=1}^{M} \int_{t_{j-1}}^{t_j} [g_{LO}(x) - l_j]^2 f(x)\, dx$$

$$= I(f) - 2 \sum_{j=1}^{M} l_j [f(t_{j-1}) - f(t_j)]$$

$$+ \sum_{j=1}^{M} l_j^2 [F(t_j) - F(t_{j-1})] \tag{4-21}$$

where $I(f)$ is Fisher's information for location shift for f defined by (2-12). From (4-21) it follows immediately that for given breakpoints the optimum choice of levels minimizing $e(g_{LO}, q; f)$ is that of (4-10). With this choice of levels the error expression of (4-21) becomes

$$e(g_{LO}, q; f) = I(f) - \tilde{\xi}_{QC,LO} \tag{4-22}$$

where $\tilde{\xi}_{QC,LO}$ is the normalized efficacy $\xi_{QC,LO}/P_e^2$ of the QC detector using these optimum or LO levels. From (4-22) it is clear that minimization of $e(g_{LO}, q; f)$ with respect to $\mathbf{t}$ is equivalent to the maximization of $\tilde{\xi}_{QC,LO}$.

The normalized efficacy $\tilde{\xi}_{QC,LO}$ may be written as

$$\tilde{\xi}_{QC,LO} = \int_{-\infty}^{\infty} q_{t,LO}(x) g_{LO}(x) f(x)\, dx \tag{4-23}$$

so that from the Schwarz inequality

$$\tilde{\xi}_{QC,LO}^2 \leq \int_{-\infty}^{\infty} q_{t,LO}^2(x) f(x)\, dx \cdot I(f)$$

$$= \tilde{\xi}_{QC,LO} I(f) \tag{4-24}$$

leading to

$$\tilde{\xi}_{QC,LO} \leq I(f) \tag{4-25}$$

This is as expected, $I(f)$ being the maximum value of the efficacy, obtained for an unconstrained AO detector, for this detection problem. We also find that for increasing M, $\tilde{\xi}_{QC,LO}$ approaches $I(f)$ for sequences of breakpoint vectors $\mathbf{t}$ which imply an increasingly finer input partitioning. In particular, this is true for the

AO partitioning satisfying (4-18) for each M.

Suppose that f is the Gaussian density function with mean μ and variance σ^2, so that $g_{LO}(x) = (x - \mu)/\sigma^2$. Let q_{MD} be the optimum quantizer minimizing the *distortion* measure $e(q;f)$ of (4-1). Then we find that the best-fit quantizer minimizing the error measure $e(g_{LO}, q;f)$ has the same breakpoints as those of q_{MD} and has output levels which are scaled and shifted versions of those of q_{MD}. Thus for Gaussian noise the quantizer giving the minimum mean-squared-error representation of the X_i under the null hypothesis is also an asymptotically optimum quantizer for the signal detection problem. But for non-Gaussian noise densities the situation can be quite different. Consider, for example, the case where f is the double-exponential noise density function of (2-15), in which case g_{LO} is the two-level characteristic given by (2-38). This is already a two-level ($M = 2$) symmetric quantizer characteristic, so that hard-limiting of the X_i is the asymptotically optimum procedure to apply in this case. The minimum distortion quantizer, on the other hand, will clearly be quite different here. For a given number of levels M the AO quantizer in general extracts in an optimum manner that information which is most relevant in detecting weak signals.

To present some numerical values let us first restrict attention to the case of symmetric noise density functions f, satisfying $f(x) = f(-x)$ for all x. It is then quite reasonable to require that the quantizer characteristic be odd-symmetric, since g_{LO} will now have odd-symmetry. We would, in fact, expect to obtain symmetric quantizers in such cases from the general solution we have obtained for the optimum quantizer parameters. It is known, however, that the minimum distortion quantizer for an input with a symmetric probability density function may not be symmetric [Abaya and Wise, 1981]. In the usual cases of interest we always get symmetry, and hence we will simply require our quantizers to be symmetric, to avoid such situations. We then need to define breakpoints and output levels only for the positive values of the inputs. Let us consider $M = 2m$ interval quantization, with m output levels for the positive inputs and m output levels for the negative inputs. Let $\mathbf{c} = (c_1, c_2, ..., c_{m-1})$ be the vector of breakpoints for the positive inputs, with $0 \triangleq c_m \leq c_{m-1} \leq \cdots \leq c_1 \leq c_0 \triangleq \infty$, and let h_j be the output level when the input is in (c_j, c_{j-1}), $j = 1,2,...,m$. To specify a $(2m - 1)$-interval quantizer with middle level of zero we can constrain h_m to be zero. The symmetric $2m$-interval quantizer characteristic $q_{\mathbf{c},\mathbf{h}}$ with breakpoints vector specified by $\mathbf{c}$ and output level vector specified by $\mathbf{h} = (h_1, h_2, ..., h_m)$ is illustrated in Figure 4.3. One relatively minor point is that at the breakpoints the quantizer output may be defined quite arbitrarily, since we are assuming that f is in general a smooth function.

For such symmetric quantization the result for $\xi_{QC,LO}$ of (4-16) becomes

$$\xi_{QC,LO} = 2P_s^2 \sum_{j=1}^{m} \frac{[f(c_j) - f(c_{j-1})]^2}{F(c_{j-1}) - F(c_j)} \qquad (4\text{-}26)$$

and the conditions (4-19) and (4-20) become, respectively,

$$g_{LO}(c_j) = \frac{h_{j+1} + h_j}{2}, \quad j = 1,2,\ldots,m-1 \qquad (4\text{-}27)$$

and

$$h_j = \frac{\int_{c_j}^{c_{j-1}} g_{LO}(x) f(x)\, dx}{F(c_{j-1}) - f(c_j)}, \quad j = 1,2,\ldots,m \qquad (4\text{-}28)$$

Equations (4-27) and (4-28) give a total of $2m - 1 = M - 1$ equations which can be solved more easily than the $2M - 1$ equations originally obtained without assuming symmetry. The asymptotically optimum breakpoints c_j may then be substituted in (4-26) to obtain the efficacy of the resulting AO quantizer detector.

Some numerical results have been obtained for the class of generalized Gaussian noise densities and for symmetric quantization [Kassam, 1977b]. Some of these results on the optimum quantizer parameters and performances are given here as Figure 4.4 and Table 4.1. Figure 4.4 shows the parameters c_1, h_1 and h_2 for four-interval ($m = 2$) quantization, both for the AO quantizer and the minimum distortion quantizer, as a function of k, the rate-of-decay parameter for the unit-variance generalized Gaussian density. It shows the significantly different characteristics of these two types of quantizers for non-Gaussian noise. Table 4.1 shows the efficacy (*detection* performance) and mean-squared error [accuracy of $q(X_i)$ as a *representation* of X_i] for both the AO and minimum mean-squared error quantizers for different values of k parameterizing the class of unit-variance generalized Gaussian densities. The significant difference in their performances based on these two criteria is quite evident, for non-Gaussian noise.

One observation which can be made from the results in Table 4.1 is that even with only four-level (or two-bit) quantization, the QC detector if properly designed can have performance which is quite good compared to that of the AO detector. While we have been considering only asymptotic performance so far, other results which have been obtained [e.g., Kassam, 1981, Kassam and Thomas, 1976] show that we can expect to obtain good relative performance in the finite sample-size case also. For example, for the Gaussian case the ARE of the AO QC detector relative to the AO

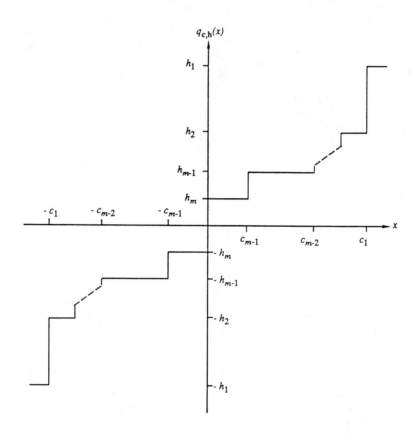

Figure 4.3 Symmetric $2m$-Interval Quantizer Characteristic

detector (the LC detector) is 0.88 (from Table 4.1). The AO value of c_1 for the QC detector is 0.98 for unit variance Gaussian noise (Figure 4.4) and the ratio h_2/h_1 for the AO QC detector for Gaussian noise is 0.3. Table 4.2, condensed from results given in [Kassam and Thomas, 1976], compares the actual detection probabilities computed for different values of θ for a four-level QC detector using approximately these AO parameter values and for the linear and sign detectors. Results for two different combinations of sample size n and false alarm probability p_f are given, for constant-signal detection. The QC detector is seen to have good relative performance, considering that it performs only a two-bit quantization of the data.

Returning to the log-likelihood ratio of (4-7) for the reduced-data matrix $\mathbf{Z}$, we note that if the signal amplitude θ_0 is given the quantity $\ln L(\mathbf{Z})$ can be obtained as a function of $\mathbf{Z}$ only. This

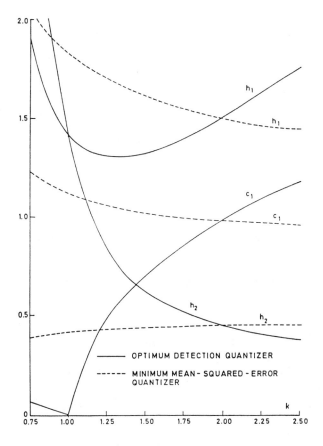

Figure 4.4 Parameters of Symmetric Four-Interval AO and
Minimum Distortion Quantizers as Functions of k for
Unit-Variance Generalized Gaussian Noise

k, Parameter of Generalized Gaussian Density	AO Quantization		Minimum MSE Quantization	
	Efficacy	MSE	Efficacy	MSE
0.75	5.267	3.787	1.802	0.213
1.00	2.000	1.000	1.214	0.176
1.25	1.285	0.418	1.034	0.153
1.50	1.033	0.218	0.955	0.138
1.75	0.926	0.141	0.912	0.126
2.00	0.883	0.117	0.883	0.117
2.25	0.870	0.121	0.860	0.111
2.50	0.876	0.140	0.842	0.105

Table 4.1 Normalized Efficacies and Mean-Squared Errors
of AO and Minimum Distortion Four-Interval Quantizers
for Unit-Variance Generalized Gaussian Noise

	Symmetric Four - Interval QC Detector $c_1 = 0.98\sigma$, $h_2/h_1 = 1/3$		Linear Detector		Sign Detector	
θ/σ	n=50	n=100	n=50	n=100	n=50	n=100
0	0.010	0.001	0.010	0.001	0.010	0.001
0.1	0.047	0.015	0.053	0.018	0.038	0.011
0.2	0.152	0.106	0.181	0.135	0.112	0.065
0.3	0.351	0.369	0.419	0.464	0.252	0.232
0.4	0.603	0.719	0.692	0.819	0.452	0.518

Table 4.2 Detection Probabilities for Symmetric Four-Interval QC Detector, Linear Detector, and Sign Detector for Constant Signal in Gaussian Noise, Variance σ^2 ($p_f = 10^{-2}$ and $n = 50$, $p_f = 10^{-3}$ and $n = 100$)

log-likelihood ratio can be interpreted as the sum of quantized X_i's, using the given breakpoints vector **t** and with *time-varying* quantizer levels which cannot in general be factored as a product of the s_i with fixed quantizer levels, as in (4-8) and (4-9). The exception to this is the constant-signal case, in which we do not get time-varying levels. For this case of constant signals and known signal amplitude, it is possible to consider the optimization of *finite* sample-size performance with respect to choice of the breakpoints vector **t**. Bushnell and Kurz [1974] have given some analysis and numerical performance results for this particular situation. For the time-varying signal case the optimization of local or asymptotic performance seems to provide the best approach to the quantizer design problem, provided the weak-signal assumption is reasonable.

In addition to the references we have given, other studies on optimum quantization for known-signal detection have been carried out by Groenveld [1972], Ching and Kurz [1972], Kurz [1977], Poor and Thomas [1977b, 1978a, 1980], Poor and Alexandrou [1980], Varshney [1981], Lee and Thomas [1983], Aazhang and Poor [1984], Ammar and Huang [1985] and Benitz and Bucklew [1986]. Among the earliest contributions to optimum quantization for weak-signal detection are those of Hansen [1969], Baronkin [1972], and Levin and Baronkin [1973]. We shall mention some further references in the rest of this chapter. Let us also note here that finite-sample-size quantizer optimization results have recently been given by Nutall [1983] and Struzinski [1983], who considered a multi-input detector with an additive signal possibly present in one of the inputs. Quantization in sequential detection of known signals has been addressed by Tantaratana and Thomas [1977]. A broader review of quantization for signal detection has been given in [Kassam, 1985].

4.3 Asymptotically Optimum Generalized Quantization

So far we have constrained ourselves to a consideration of multilevel quantization of the input data with the usual interpretation that quantization involves partitioning the range of real values of the input into some number M of *intervals*. We may describe this operation more explicitly as *M-interval partitioning*, with an output level assigned to each input interval. Such an M-interval partitioning is specified by a vector **t** of breakpoints, or interval end points.

Now for any given vector **t**, it should be clear that the optimum choice of levels l_j need not result in M distinct output levels. Specifically, for given **t** the levels specified by (4-10), which are the LO levels and give the best fit to g_{LO}, need not be distinct. Figure 4.5 illustrates a non-trivial situation in which this can happen, with g_{LO} itself not a piecewise constant function. In particular, g_{LO} has the characteristic shown in Figure 4.5 for the Cauchy noise density function. In a *generalized* sense we can think of the quantization depicted in Figure 4.5 as a *four-level* quantization, with four *regions* partitioning the range of input values and a *distinct* output level associated with each region.

Ordinary four-interval quantization (using four levels) in the situation of Figure 4.5 does not utilize optimally the availability of four distinct levels in approximating g_{LO}, because the regions of the input partitioning are *constrained* to be intervals. It is clear that M-level generalized quantization allows optimum use of distinct levels. Note that we may refer to M-level generalized quantization as *M-region quantization* based on *M-region partitioning*. Since it is the number of distinct output levels or regions of the input partitioning which determines the data-compression value of a quantizer, it is appropriate for us to extend our analysis to M-region quantization.

Let $\{A_j, j = 1,2,...,M\}$ denote the partitioning of the real line $\mathbb{R} = (-\infty,\infty)$ into disjoint subsets A_j for an M-region quantization. For a *given* partitioning, it is easy to obtain the LO levels for the M-region quantization. We note simply that Z_{ij} of (4-4) is now defined as

$$Z_{ij} = \begin{cases} 1, & X_i \in A_j \\ 0, & \text{otherwise} \end{cases} \quad j = 1,2,...,M \quad (4\text{-}29)$$

and $p_{ij}(\theta)$ of (4-6) becomes

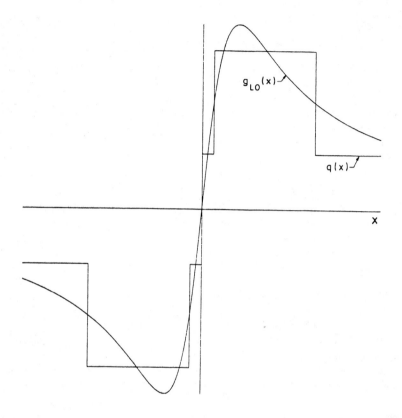

Figure 4.5 Generalized Quantization of Data

$$p_{ij}(\theta) = P\{X_i \in A_j\}$$
$$= \int_{A_j} f(x - \theta s_i)\, dx \qquad (4\text{-}30)$$

From this we get easily, following the development for the LO levels of (4-10), that the LO levels for M-region quantization are given by

$$l_j = \frac{\int_{A_j} -f'(x)\, dx}{\int_{A_j} f(x)\, dx}$$

$$= \frac{\int_{A_j} g_{LO}(x) f(x)\, dx}{\int_{A_j} f(x)\, dx}, \quad j = 1, 2, \ldots, M \qquad (4\text{-}31)$$

This result should be compared with that of (4-20), where $A_j = (t_{j-1}, t_j]$.

In the same way, it is easy to establish counterparts for the efficacies ξ_{QC} and $\xi_{QC,LO}$, which are

$$\xi_{QC} = P_s^2 \frac{\left[\sum_{j=1}^{M} (l_j - l_0) \int_{A_j} g_{LO}(x) f(x) \, dx\right]^2}{\sum_{j=1}^{M} (l_j - l_0)^2 \int_{A_j} f(x) \, dx} \qquad (4\text{-}32)$$

and

$$\xi_{QC,LO} = P_s^2 \sum_{j=1}^{M} \frac{\left[\int_{A_j} g_{LO}(x) f(x) \, dx\right]^2}{\int_{A_j} f(x) \, dx} \qquad (4\text{-}33)$$

In addition, the mean-squared error $e(g_{LO}, q; f)$ now becomes

$$e(g_{LO}, q; f) = I(f) - 2 \sum_{j=1}^{M} l_j \int_{A_j} g_{LO}(x) f(x) \, dx$$
$$+ \sum_{j=1}^{M} l_j^2 \int_{A_j} f(x) \, dx \qquad (4\text{-}34)$$

The LO levels of (4-31) are also those minimizing $e(g_{LO}, q; f)$ for given subsets A_j of the input partitioning.

The problem of finding the asymptotically optimum regions A_j maximizing $\xi_{QC,LO}$ of (4-33) is not as simple as that of finding the AO breakpoints for M-interval quantization. It is, nonetheless, possible to find a general result for the AO regions. For this it is expedient to view our problem as that of picking the M-region quantizer which gives the minimum value for $e(g_{LO}, q; f)$, since we have observed that this will also give us the AO quantizer. Now we can write $e(g_{LO}, q; f)$ as

$$e(g_{LO}, q; f) = \sum_{j=1}^{M} \int_{A_j} (g_{LO}(x) - l_j)^2 \, dF(x)$$
$$= \sum_{j=1}^{M} \int_{-\infty}^{\infty} p_j(x) I_j(x) \, dF(x) \qquad (4\text{-}35)$$

where

$$p_j(x) = (g_{LO}(x) - l_j)^2 \qquad (4\text{-}36)$$

and I_j is the indicator function

$$I_j(x) = \begin{cases} 1, & x \in A_j \\ 0, & \text{otherwise} \end{cases} \quad (4\text{-}37)$$

Let l_j, $j = 1,2,...,M$, be a *given* set of distinct levels, and assume without any loss of generality that we have $l_1 < l_2 < \cdots < l_M$. Now it is obvious that for $e(g_{LO}, q; f)$ to be a minimum, the regions A_j must be such as to contain the subsets $\tilde{A}_j$ defined by

$$\tilde{A}_j = \{x \mid p_j(x) < p_i(x), \text{ all } i \neq j\} \quad (4\text{-}38)$$

for $j = 1,2,...,M$. Since $p_j(x) = g_{LO}^2(x) - 2l_j g_{LO}(x) + l_j^2$, we find that the $\tilde{A}_j$ are

$$\tilde{A}_j = \{x \mid 2(l_i - l_j) g_{LO}(x) < l_i^2 - l_j^2, \text{ all } i \neq j\} \quad (4\text{-}39)$$

Defining

$$\tilde{t}_j \triangleq \frac{l_j + l_{j+1}}{2}, \quad j = 1,2,...,M-1 \quad (4\text{-}40)$$

we find that $\tilde{t}_1 < \tilde{t}_2 < \cdots < \tilde{t}_{M-1}$, so that by considering the cases $i > j$ and $i < j$ separately in (4-39) we get the result

$$\tilde{A}_j = \{x \mid \tilde{t}_{j-1} < g_{LO}(x) < \tilde{t}_j\} \quad (4\text{-}41)$$

with $\tilde{t}_0 \triangleq -\infty$ and $\tilde{t}_M \triangleq \infty$. Assuming that the set of points at which $g_{LO}(x)$ is equal to the $\tilde{t}_j$, $j = 1,2,...,M$, has probability zero, we finally get

$$A_j = \{x \mid \tilde{t}_{j-1} < g_{LO}(x) \leq \tilde{t}_j\}, \quad j = 1,2,...,M \quad (4\text{-}42)$$

Note carefully that A_j above are the optimum regions giving the quantizer best fitting g_{LO} under H_1 for a given set of levels l_j. We *cannot* conclude that they are also the AO choice of regions maximizing ξ_{QC} for the *given* levels. However, the set of $2M - 1$ *simultaneous* equations (4-31) and (4-40) together with (4-42) for the levels and regions are the necessary conditions for maximization of ξ_{QC}.

With A_j defined according to (4-42) and the $\tilde{t}_j$ any increasing sequence of constants, the levels l_j of (4-31) will turn out to form an increasing sequence, for $j = 1,2,...,M$. If g_{LO} is a monotonic function the optimum regions will clearly turn out to be intervals, and a restriction to M-level quantizers based on M-

interval partitioning is justified. Otherwise, for a given M the AO M-region quantizer will have better asymptotic performance than the AO M-interval quantizer. Recently, Wong and Blake [1986] have described an interesting algorithm using independent noise samples which can be used to design optimum or almost optimum generalized quantizers.

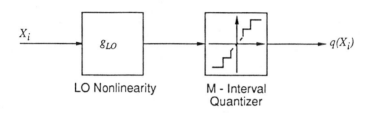

Figure 4.6 Asymptotically Optimum M-Level Quantization for Known-Signal Detection

We can give an interesting interpretation to the above AO M-region quantization scheme. The quantizer can be implemented as shown in Figure 4.6, with each X_i first transformed into $g_{LO}(X_i)$, which then forms the input to an M-interval quantizer with breakpoints vector $\tilde{\mathbf{t}} = (\tilde{t}_1, \tilde{t}_2, ..., \tilde{t}_{M-1})$ and output levels $l_1, l_2, ..., l_M$. We have, of course, the relationship (4-40) between the $\tilde{t}_j$ and the levels l_j, so that the M-interval quantizer in Figure 4.6 satisfies the condition that the output level l_j for the j-th interval $(\tilde{t}_{j-1}, \tilde{t}_j)$ lies in this interval. This suggests a suboptimum scheme in which the M-interval quantizer in Figure 4.6 is replaced by a *uniform* quantizer. The resulting scheme may then be optimized by picking the uniform quantizer step size and saturation levels optimally for given M. It is interesting to note that Poor and Rivani [1981] arrived at such a scheme, based on a uniform quantization for the LO transformed data $g_{LO}(X_i)$, as a special case of a more general system employing an odd nonlinear function g before a uniform quantizer (for the case of symmetric f and symmetric quantization).

4.4 Maximum-Distance Quantization

For a given vector $\mathbf{t}$ of breakpoints for an M-interval quantizer, the levels l_j of (4-10) lead to a locally optimum detector based on M-interval partitioned data. To further optimize the design of the quantizer we obtained the asymptotically optimum set of breakpoints. This was done because it is not easy to obtain general results for locally optimum breakpoints for finite sample sizes n. The reason why we considered the weak-signal situation in the first place is that the log-likelihood ratio for detection based on inputs partitioned by $\mathbf{t}$ depends on the signal amplitude in an essential way, as can be seen from (4-7) and (4-6). Suppose that we did have a particular value $\theta = \theta_0 > 0$ for the amplitude of the known signal, when it is present. Then the log-likelihood ratio of (4-7) could be used to obtain an optimum detection scheme, and would result in the optimum quantizer having *time-varying* levels. But even in this case, further optimization with respect to $\mathbf{t}$ is quite difficult to perform. For this one would have to find the detection probability of the optimum detector using the log-likelihood ratio $\ln L(\mathbf{Z})$ as a test statistic, and then obtain the breakpoints vector maximizing the detection probability (for the given values of n and θ_0 and the noise density function f). The fact that the expression for detection probability does not lend itself to analytical optimization techniques, and depends on p_f, makes this approach applicable only in special cases and does not allow easy generation of general results useful for design.

As an alternative to these approaches for the optimum choice of quantizer breakpoints, Poor and Thomas [1977a] suggested a design procedure based on the maximization of a *distance measure*. The basic concept here may be described as follows. Our objective in M-interval partitioning is to obtain breakpoints t_j leading to the "best" reduced data matrix $\mathbf{Z}$, which has elements defined by (4-4). Now for given $\mathbf{t}$ the elements of $\mathbf{Z}$ have a joint density function $p_\mathbf{Z}(\mathbf{z} \mid \theta)$. Since our detection of signals is based on the use of $\mathbf{Z}$, we would like $\mathbf{Z}$ to have characteristics under the conditions $\theta = 0$ and $\theta = \theta_0$ which are as *different* as possible, so that observation of $\mathbf{Z}$ can give a good indication as to whether $\theta = 0$ or $\theta = \theta_0$ is the true condition. Thus we would like to have $p_\mathbf{Z}(\mathbf{z} \mid 0)$ and $p_\mathbf{Z}(\mathbf{z} \mid \theta_0)$ be as different as possible, or to have maximum distance between them, according to some reasonable measure of distance or separation between two probability density functions.

The use of measures of distance between two probability distributions has in the past been quite fruitful in signal processing applications; Grettenberg [1963] and Kailath [1967] give, for example, several applications and properties of some useful distance measures. An interesting class of distance measures between two probability distributions is the class of *Ali-Silvey distances*, which generates many of the common distance measures which have been used in various applications. In terms of our specific situation, an Ali-Silvey distance $d_\mathbf{Z}$ between the density functions $p_\mathbf{Z}(\mathbf{z} \mid 0)$ and $p_\mathbf{Z}(\mathbf{z} \mid \theta_0)$ is any distance measure of the form

$$d_Z = E\{C[L(Z)] \mid \theta = 0\} \tag{4-43}$$

where $L(Z)$ is the likelihood ratio

$$L(Z) = \frac{p_Z(Z \mid \theta_0)}{p_Z(Z \mid 0)} \tag{4-44}$$

and C is any convex function on $(0,\infty)$. Two common distance measures which are Ali-Silvey distances (or increasing functions of Ali-Silvey distances) are the *J-divergence* and the *Bhattacharyya distance*. The J-divergence is the distance

$$d_Z^J = E\{[L(Z) - 1] \ln L(Z) \mid \theta = 0\} \tag{4-45}$$

which is an Ali-Silvey distance with $C(x) = (x-1)\ln x$. The Bhattacharyya distance d_Z^B for the densities $p_Z(z \mid 0)$ and $p_Z(z \mid \theta_0)$ is

$$d_Z^B = -\ln(1 - d^2) \tag{4-46}$$

where

$$d^2 = E\{[\sqrt{L(Z)} - 1]^2 \mid \theta = 0\} \tag{4-47}$$

Thus d^2 is an Ali-Silvey distance with $C(x) = (\sqrt{x} - 1)^2$, and d_Z^B is an increasing function of an Ali-Silvey distance.

The utility of distances such as the J-divergence and the Bhattacharyya distance is based on several results showing the close relationship between probability of error and the distance measure in hypothesis-testing problems. In addition to the references just cited, further discussion on this topic can be found in [Ali and Silvey, 1966], [Kobayashi, 1970], and [Kobayashi and Thomas, 1967].

Let us consider the *constant-signal* case, and furthermore consider the Ali-Silvey distance between the H_1 and K_1 distributions of each *individual* quantizer output $q(X_i)$. With θs_i now taken to be some known value θ_0 for all i, we find that (using the notation of Section 4.2)

$$L(Z_i) = \prod_{j=1}^{M} \left[\frac{F(t_j - \theta_0) - F(t_{j-1} - \theta_0)}{F(t_j) - F(t_{j-1})}\right]^{Z_{ij}} \tag{4-48}$$

Defining $\Delta_j(\theta)$ as

$$\Delta_j(\theta) \triangleq F(t_j - \theta) - F(t_{j-1} - \theta) \qquad (4\text{-}49)$$

we get from the definition (4-43)

$$d_{Z_i} = \sum_{j=1}^{M} C\left[\frac{\Delta_j(\theta_0)}{\Delta_j(0)}\right] \Delta_j(0) \qquad (4\text{-}50)$$

For maximization of d_{Z_i} we take the partial derivatives of d_{Z_i} with respect to the t_j and set them equal to zero to obtain necessary conditions. This procedure gives a set of simultaneous equations which can be solved numerically. Of course the solution will depend on the choice of C, which is one drawback of this approach. Poor and Thomas [1977a] have shown, however, that in the limiting case $\theta_0 \to 0$, the results do become those we have established for asymptotically optimum breakpoints [Equation (4-18)], for any choice of C.

The main advantage of this approach is that one does not have to use the detection probability directly in the optimization. There are, however, limitations, some of which we have noted above. For example, even in the constant-signal case, use of d_Z instead of d_{Z_i} will generally lead to different results. One exception to this is the case of J-divergence, for which we get the same design for **t** using d_Z or d_{Z_i} (Problem 4.6). For the time-varying-signal case one should maximize d_Z with respect to **t** for a fixed quantizer design, and this can also lead to rather complicated equations. Thus, although the use of distance measures does provide a useful alternative for quantizer design in the constant-signal case, for time-varying signals its use is less easy. Nonetheless, in this situation it might still be easier to maximize the distance d_Z rather than a detection probability expression directly. One possible approach in the case of time-varying, non-vanishing signals is to design the quantizer for a constant signal with the same average power.

4.5 Approximations of Locally Optimum Test Statistics

In this last section we extend the ideas of the above sections on AO quantization by considering approximations to the LO nonlinearity from more general classes of allowable functions, which include the classes of piecewise-constant quantizers as special cases.

The locally optimum test statistic for our known signal detection problem is given by $\lambda_{LO}(\mathbf{X})$ of (2-32),

$$\lambda_{LO}(\mathbf{X}) = \sum_{i=1}^{n} s_i \, g_{LO}(X_i) \qquad (4\text{-}51)$$

If $\lambda_{LO}(\mathbf{X})$ cannot be implemented exactly, a sub-optimum statistic has to be used. Let us now consider the use of some GC statistic

$$T_{GC}(\mathbf{X}) = \sum_{i=1}^{n} a_i \, g(X_i) \qquad (4\text{-}52)$$

in place of $\lambda_{LO}(\mathbf{X})$, where the coefficients a_i and the characteristic g have to be chosen from some constrained set of possibilities. For example, as we did in Section 4.2, g may be constrained to be an M-interval quantizer characteristic. Here we will generalize the situation and assume that there is some class $\mathbf{G}$ of characteristics from which the best one has to be picked. In addition, the coefficients a_i may be allowed only a finite number of bits in implementation and may be subject to other constraints (for example, the a_i may be constrained to be powers of 2 for multiplication-free implementation). Thus in general we will assume that there is some set of allowable coefficient n-vectors $\mathbf{A}_n$ from which the best $\mathbf{a} = (a_1, a_2, ..., a_n)$ has to be chosen.

Optimum Choice of Characteristic g

At first let us concentrate on the problem of choosing g in $\mathbf{G}$ optimally, and assume that we have $\mathbf{a} = \mathbf{s}$. Now the normalized efficacy $\tilde{\xi}_{GC}$ of any GC detector with $\mathbf{a} = \mathbf{s}$ is

$$\tilde{\xi}_{GC} = \frac{\left[\int_{-\infty}^{\infty} g(x) f'(x) \, dx\right]^2}{\int_{-\infty}^{\infty} g^2(x) f(x) \, dx}$$

$$= \frac{\left[\int_{-\infty}^{\infty} g(x) g_{LO}(x) f(x) \, dx\right]^2}{\int_{-\infty}^{\infty} g^2(x) f(x) \, dx} \qquad (4\text{-}53)$$

Adopting $\tilde{\xi}_{GC}$ as our performance criterion, we want to characterize the optimum choice of nonlinearity g in $\mathbf{G}$ which maximizes performance. Note that in defining GC detectors we make assumption D (iii) [(2-58) and (2-59)], so that here too $\mathbf{G}$ will be assumed to contain functions which satisfy these constraints. We also know that $\tilde{\xi}_{GC}$ is invariant to amplitude scaling of the nonlinear characteristic g; thus we are justified in considering instead of the set $\mathbf{G}$ the larger set $\overline{\mathbf{G}}$ of functions which are all the variously amplitude scaled versions of functions g in $\mathbf{G}$. Each member of $\overline{\mathbf{G}}$ is associated with a member of $\mathbf{G}$ and an amplitude scale parameter c.

In Section 4.2 we saw that if $\overline{\mathbf{G}}$ is the class of all M-interval quantizers, an optimum quantizer in $\overline{\mathbf{G}}$ can be interpreted as being the minimum mean-squared-error fit to g_{LO} under f. We now show that such a characterization holds in the more general case. In the notation of Section 4.2, $e(g_{LO}, \overline{g}; f)$ is the mean-squared error

$$e(g_{LO}, \overline{g}; f) = \int_{-\infty}^{\infty} [g_{LO}(x) - cg(x)]^2 f(x)\, dx \qquad (4\text{-}54)$$

where $\overline{g}$ in $\overline{\mathbf{G}}$ is defined in terms of c and some g in $\mathbf{G}$. This mean-squared error may be expressed as

$$e(g_{LO}, \overline{g}; f) = I(f) - 2c \int_{-\infty}^{\infty} g(x) g_{LO}(x) f(x)\, dx$$
$$+ c^2 \int_{-\infty}^{\infty} g^2(x) f(x)\, dx \qquad (4\text{-}55)$$

In order to minimize $e(g_{LO}, \overline{g}; f)$ with respect to $\overline{g}$ in the set $\overline{\mathbf{G}}$, let us first consider for fixed g in $\mathbf{G}$ minimization with respect to c. Differentiating with respect to c and setting the result equal to zero gives the minimization condition

$$c = \frac{\int_{-\infty}^{\infty} g(x) g_{LO}(x) f(x)\, dx}{\int_{-\infty}^{\infty} g^2(x) f(x)\, dx} \qquad (4\text{-}56)$$

With this value for c in (4-55), the mean-squared error becomes

$$e(g_{LO}, g; f) = I(f) - \frac{\left[\int_{-\infty}^{\infty} g(x) g_{LO}(x) f(x)\, dx\right]^2}{\int_{-\infty}^{\infty} g^2(x) f(x)\, dx}$$

$$= I(f) - \tilde{\xi}_{GC} \qquad (4\text{-}57)$$

Thus the maximization of $\tilde{\xi}_{GC}$ with respect to g in $\mathbf{G}$ is equivalent to the minimization of $e(g_{LO}, \bar{g}; f)$ with respect to $\bar{g}$ in $\overline{\mathbf{G}}$. To find the best sub-optimum nonlinear characteristic in $\mathbf{G}$, one can therefore seek that scale constant c and that g in $\mathbf{G}$ which gives the *best fit* to g_{LO} in the mean-squared-error sense under f. Our derivation of this result is similar to one given by Czarnecki and Vastola [1985]. This general result allows us to compare two non-linear characteristics g_1 and g_2 for $T_{GC}(\mathbf{X})$ on an intuitive basis; if g_1 appears to provide a better fit to g_{LO} than does g_2 on the average under noise-only conditions, then g_1 will give a more efficient test statistic. Further considerations of this point may be found in [Halverson and Wise, 1986].

One important special case occurs when $\mathbf{G}$ is specified to be all linear combinations of some finite number of functions $g_1, g_2, ..., g_M$. Then $\mathbf{G}$ and $\overline{\mathbf{G}}$ are the same and any g in $\mathbf{G}$ is expressed as

$$g(x) = \sum_{j=1}^{M} l_j g_j(x) \tag{4-58}$$

for some set of constants $l_1, l_2, ..., l_M$. In particular, suppose that the set of functions $\{g_1, g_2, ..., g_M\}$ is an *orthonormal* set, so that

$$\int_{-\infty}^{\infty} g_j(x) g_k(x) f(x)\, dx = \delta_{jk} \tag{4-59}$$

where δ_{jk} is the Kronecker delta function. Then the mean-squared error $e(g_{LO}, g; f)$ becomes

$$e(g_{LO}, g; f) = I(f) - 2\sum_{j=1}^{M} l_j <g_j, g_{LO}>_f \\ + \sum_{j=1}^{M} l_j^2 \tag{4-60}$$

and the optimum weights are

$$l_j = <g_j, g_{LO}>_f \tag{4-61}$$

where the inner products are defined by

$$<g, h>_f = \int_{-\infty}^{\infty} g(x) h(x) f(x)\, dx$$

$$= E\{g(X) h(X) \mid f\} \tag{4-62}$$

Notice that all M-interval quantizers with a fixed breakpoints vector $\mathbf{t}$ can be represented in the form of (4-58) with orthonormal basis functions g_j. In the general case the optimum weights l_j are simply the projections of g_{LO} on the individual g_j.

Adaptive Detection

Suppose the set of functions $\{g_j\}_{j=1}^M$ is known to be orthogonal for some collection $\overline{\mathbf{F}}$ of *unit-variance* versions $\overline{f}$ of noise density functions f which are possible in a given situation. This will be true, in particular, when the g_j are non-zero over respective sets P_j which are disjoint. The set of functions g_j can be picked to provide a set of basis functions which are useful in approximating the LO characteristic $\overline{g}_{LO}$ for any density $\overline{f}$ in $\overline{\mathbf{F}}$. Assuming that any adaptive scheme will also obtain an estimate $\hat{\sigma}$ of the noise standard deviation σ, we may restrict attention to the class $\overline{\mathbf{F}}$ of densities for the normalized observations X_i/σ under H_1.

For a GC test statistic based on the linear combination of (4-58), the optimum weights are now given by

$$l_j = \frac{<g_j, \overline{g}_{LO}>_{\overline{f}}}{<g_j, g_j>_{\overline{f}}} \tag{4-63}$$

This is a slight generalization of the result (4-61), since we have assumed only that the set $\{g_j\}_{j=1}^M$ of functions is an orthogonal set for all $\overline{f}$ in $\overline{\mathbf{F}}$. Note that

$$<g_j, \overline{g}_{LO}>_{\overline{f}} = \int_{-\infty}^{\infty} g_j(x) \overline{g}_{LO}(x) \overline{f}(x) \, dx$$

$$= \sigma \int_{-\infty}^{\infty} g_j(x/\sigma) g_{LO}(x) f(x) \, dx$$

$$= -\sigma \int_{-\infty}^{\infty} g_j(x/\sigma) f'(x) \, dx \tag{4-64}$$

and

$$<g_j, g_j>_{\overline{f}} = \int_{-\infty}^{\infty} g_j^2(x) \overline{f}(x) \, dx$$

$$= \int_{-\infty}^{\infty} g_j^2(x/\sigma) f(x) \, dx \tag{4-65}$$

so that the optimum levels are

$$l_j = \frac{-\sigma \int_{-\infty}^{\infty} g_j(x/\sigma) f'(x) \, dx}{\int_{-\infty}^{\infty} g_j^2(x/\sigma) f(x) \, dx} \tag{4-66}$$

Assuming that σ can be estimated, the denominator for l_j above may be obtained as the estimate of the expected value of $g_j^2(X_i/\sigma)$ under noise-only conditions. If in addition a suitable adaptive scheme can be formulated to estimate the numerator term, then a useful adaptive detector can be implemented. The fact that the functions g_j are orthogonal for the collection $\overline{\mathbf{F}}$ makes the estimation of the l_j a much easier task, allowing complete decoupling of the estimation for different l_j.

A key requirement above is that $\{g_j\}_{j=1}^M$ be an orthogonal set of functions for all $\overline{f}$ in $\overline{\mathbf{F}}$. As we have mentioned, this is true in particular when the g_j have disjoint supports P_j. An important special case is that of quantization, for which we have

$$g_j(x) = \begin{cases} 1, & t_{j-1} < x \leq t_j \\ 0, & \text{otherwise} \end{cases} \tag{4-67}$$

for some given vector $\mathbf{t} = (t_0, t_1, ..., t_M)$ of non-decreasing components with $t_0 = -\infty$ and $t_M = \infty$. In this special case (4-66) gives

$$l_j = \sigma \frac{f(t_{j-1}\sigma) - f(t_j\sigma)}{F(t_j\sigma) - F(t_{j-1}\sigma)} \tag{4-68}$$

For adaptive quantization we need to estimate σ as well as the values of f and F at the σt_j, $j = 1,2,...,M-1$. Of course here we have a fixed breakpoints vector $\mathbf{t}$, chosen to allow a good approximation to the LO nonlinearities corresponding to the class of densities $\overline{\mathbf{F}}$. This explains why scaling by σ is necessary in (4-68), compared with the result of (4-10).

For some further details of such detection schemes based on the use of orthogonal approximating functions, we refer the reader to [Mihajlovic and Kurz, 1982].

Optimum Choice of Coefficients

Although we assumed at the beginning that $\mathbf{a} = \mathbf{s}$, it should be clear from the general efficacy expression

$$\xi_{GC} = \rho_{as}^2 P_s^2 E^2(g,f) \tag{4-69}$$

[see (2-73)] that the optimization with respect to g in $\mathbf{G}$ of $T_{GC}(\mathbf{X})$ is decoupled from the optimization with respect to $\mathbf{a}$, for this *asymptotic* criterion. One problem that arises when considering optimization of the coefficient vector $\mathbf{a}$ in $\mathbf{A}_n$ is that for finite n (and nonzero θ) the performance of a GC detector cannot generally be characterized by a mathematical expression which is analytically convenient. If, however, we make the assumption that n is reasonably large we may use the approximation

$$\rho_{as}^2 P_s^2 \approx \frac{1}{n} \frac{\left[\sum_{i=1}^{n} a_i s_i\right]^2}{\sum_{i=1}^{n} a_i^2}$$

$$\triangleq d^2(\mathbf{a},\mathbf{s}) \tag{4-70}$$

in (4-69), in light of the definitions (2-44), (2-57), (2-64), and (2-68). What this means is that in the result for the approximate detection probability of (2-70) we now have the approximation $P_s \rho_{as} \approx d(\mathbf{a},\mathbf{s})$. Thus it is reasonable for large n to choose $\mathbf{a}$ to maximize $d^2(\mathbf{a},\mathbf{s})$. For a given finite set $\mathbf{A}_n$ it is possible to seek the best coefficient vector $\mathbf{a}$ through an exhaustive search, although this approach may not be feasible for typical values of n. For particular types of constrained sets $\mathbf{A}_n$ it may be possible to design efficient search algorithms. An example of such an algorithm for finite-bit coefficients has been given by Kassam and Lim [1978].

It is also possible to characterize, as in the case of optimization over $\mathbf{G}$ of $\tilde{\xi}_{GC}$, the optimum vector in $\mathbf{A}_n$ maximizing $d^2(\mathbf{a},\mathbf{s})$ as a scaled version of the vector in $\overline{\mathbf{A}}_n$ "closest" to $\mathbf{s}$. Here $\overline{\mathbf{A}}_n$ contains all scaled versions of vectors in $\mathbf{A}_n$. Let us also mention that the optimization of coefficient quantization has been considered for the case where the X_i are dependent, by Chen and Kassam [1983]. Here, as one can expect, the "best-fit" coefficient vector does not necessarily produce the optimum choice; the result now depends on the covariance matrix of $[g(X_1), g(X_2),...,g(X_n)]$.

PROBLEMS

Problem 4.1

Verify that for a given breakpoints vector $\mathbf{t}$, the efficacy ξ_{QC} of (4-14) is maximized by the levels l_j of (4-10).

Problem 4.2

Show that the conditions (4-18) are necessary for maximization of the efficacy $\xi_{QC,LO}$ of (4-16).

Problem 4.3

Obtain the conditions that have to be satisfied by the AO breakpoints vector **t** maximizing ξ_{QC} of (4-14) for a *given* set of levels l_j, $j = 1,2,...,M$. Simplify your result for *symmetric* $2m$ level quantization and symmetric f. What are the AO breakpoints for $m = 2$ and levels $-1,0,0,1$? Compare the result for Gaussian f with the result of Problem 2.6 in Chapter 2.

Problem 4.4

A QC detector is based on a *uniform* symmetric $2m$-interval partitioning of the input range, so that its breakpoints in $[0,\infty]$ are $c_j = (1 - j/m)v$, $j = 1,...,m$. For f a pdf which is symmetric about the origin in our known-signal detection problem, obtain the AO design equations for the overload point v and the output levels h_j. Compare the efficacies of the AO uniform-partitioning QC detector and the AO QC detector for f in the class of generalized Gaussian pdf's.

Problem 4.5

A GC detector for the known-signal detection problem uses *uniformly* quantized values of *transformed* observations $h(X_i)$, $i = 1,2,...,n$. The uniform quantization is based on a symmetric $2m$-interval partitioning as in Problem 4.4, the output level for quantizer input in (c_j, c_{j-1}) being $(m - j + 1/2)\delta$ for $j = 1,2,...,m$, with $c_0 \triangleq \infty$. The pdf f of the noise is assumed to be strictly positive, symmetric about the origin, and strongly unimodal ($-\ln f$ is strictly convex), with f' continuous. The transformation h is odd-symmetric, continuous, and strictly increasing.

(a) Obtain an expression for the normalized efficacy of such a GC detector. Find the optimum function h and the overload point v maximizing this efficacy.

(b) For f the logistic pdf of Problem 3.3 in Chapter 3, compute v and determine the function h when $m = 2$. Compute the resulting value of the efficacy and compare it with that of the corresponding AO GC detector.

(c) Repeat (b) for f the Gaussian pdf.

Problem 4.6

Show that for detection of a constant signal with a known value, the design of a maximum-distance quantizer based on the J-divergence d_Z^J is independent of the sample size n.

Problem 4.7

Let g in a GC detector be of the form
$$g(x) = \gamma x + (1 - \gamma) \operatorname{sgn} x$$
This characteristic was considered in Section 3.4 of Chapter 3 for adaptive detection. Consider amplitude-scaled versions cg of g. Show that the value of γ for which cg is the closest to g_{LO} in the mean-squared-error sense is the value specified by (3-31).

Chapter 5

DETECTION OF KNOWN NARROWBAND SIGNALS IN NARROWBAND NOISE

5.1 Introduction

In the last three chapters we have been concerned with the detection of known signals which may be characterized as being *low-pass* signals. We did not put any explicit constraints on the known signal values s_i, but required the noise components W_i to be independent for $i = 1,2,...,n$. When the data are generated as samples of some continuous-time input process, the values s_i are samples of a signal waveform $s(t)$. Thus a usually implicit assumption in this situation is that the highest signal frequency F_s is smaller than the bandwidth F of the frequency band $[0,F]$ on which the noise power spectral density can be assumed to be approximately constant.

We will now consider the situation where the underlying continuous-time signal is a *narrowband* signal and the noise process is *also* a narrowband process. A narrowband deterministic signal is an RF carrier with low-frequency deterministic amplitude and phase modulations. In this chapter the signal is still assumed to be completely known, so that the carrier frequency f_0 and carrier phase, as well as the amplitude and phase modulating functions, are assumed to be known.

Suppose the noise process is not narrowband but has a power spectral density function which is approximately constant over a range $[0,F]$ of frequencies, and suppose the narrowband signal bandwidth B_s (for signal band centered on f_0) is such that $f_0 + B_s/2 \leq F$. In this case no new results are necessary, the signal being of a low-pass type *relative* to the noise; the use of a sampling frequency of $2F$ will result in the noise components being approximately uncorrelated (which would be taken to imply independence), and will result in signal samples from which the continuous-time narrowband signal process can be reconstructed. Note that the signal being narrowband, we have $B_s \ll f_0$. Now suppose the noise is also a narrowband random process, with bandwidth $B \geq B_s$ for a frequency band centered on f_0. If the noise process has an approximately flat spectral density on $[f_0 - B/2, f_0 + B/2]$, one can use a sampling rate of B and obtain approximately uncorrelated noise samples. Indeed, over an observation interval of length T one can generate $2BT$ uncorrelated samples, by sampling both the in-phase and quadrature components of the observed waveform at a rate of B. Thus essentially all of the signal amplitude and phase modulation can be utilized in forming the detection statistic. This requires, however, that we now perform a new analysis for detection based on in-phase and quadrature observation components which may *not* be independent at any particular sampling instant.

Many of the results in this chapter can be attributed to the works of two groups of investigators, namely Zachepitsky, Mareskin, and Pakhomov [1972], Modestino [1975] and Modestino and Ningo [1979]. In developing these results here we will follow the lines of our more detailed development of corresponding results for detection of low-pass signals.

5.2 The Observation Model

In the continuous-time domain the general representation of a completely known narrowband signal waveform observed in additive noise is

$$X(t) = \theta v(t) \cos[\omega_0 t + \phi(t)] + W(t) \tag{5-1}$$

Here $v(t)$ and $\phi(t)$ are known amplitude and phase modulations which occupy a narrow band of frequencies around the carrier frequency $f_0 = \omega_0/2\pi$ and θ is, as before, the overall signal amplitude. The noise process $W(t)$ will be assumed to be stationary, zero-mean, bandpass white noise, with a power spectral density which is a constant, $N_0/2$, over a band of frequencies $(f_0 - B/2, f_0 + B/2)$ and zero outside. The signal bandwidth B_s will be assumed to be no larger than B. We are interested, of course, in testing $\theta = 0$ versus $\theta > 0$.

The noise process $W(t)$ can be expressed in terms of its low-pass in-phase and quadrature components $W_I(t)$ and $W_Q(t)$, respectively, as

$$W(t) = W_I(t) \cos \omega_0 t + W_Q(t) \sin \omega_0 t \tag{5-2}$$

The components $W_I(t)$ and $W_Q(t)$ can be generated by passing the product of $W(t)$ with $2\cos\omega_0 t$ and $2\sin\omega_0 t$, respectively, through an ideal low-pass filter of bandwidth B. Expanding the signal term $v(t)\cos[\omega_0 t + \phi(t)]$ as $v(t)\cos\phi(t)\cos\omega_0 t - v(t)\sin\phi(t)\sin\omega_0 t$, we get the representation

$$X(t) = [\theta s_I(t) + W_I(t)] \cos \omega_0 t$$
$$+ [\theta s_Q(t) + W_Q(t)] \sin \omega_0 t \tag{5-3}$$

where

$$s_I(t) = v(t) \cos \phi(t) \tag{5-4}$$

and

$$s_Q(t) = -v(t)\sin\phi(t) \qquad (5\text{-}5)$$

Defining the in-phase and quadrature observation components as

$$X_I(t) = \theta s_I(t) + W_I(t) \qquad (5\text{-}6)$$

and

$$X_Q(t) = \theta s_Q(t) + W_Q(t) \qquad (5\text{-}7)$$

we get the representation

$$X(t) = X_I(t)\cos\omega_0 t + X_Q(t)\sin\omega_0 t \qquad (5\text{-}8)$$

Note that these in-phase and quadrature components can be physically generated at the receiver by simple modulation and low-pass filtering operations. Assuming, then, that these components are available over some time interval $[T_1, T_2]$, we can sample their waveforms at a uniform rate of $1/\Delta$ to obtain a set of in-phase observations X_{Ii}, $i = 1,2,...,n$, and a set of quadrature observations X_{Qi}, $i = 1,2,...,n$, where $n = (T_2 - T_1)/\Delta$:

$$\begin{aligned} X_{Ii} &= X_I(i\Delta) \\ &= \theta s_I(i\Delta) + W_I(i\Delta) \\ &\triangleq \theta s_{Ii} + W_{Ii}, \quad i = 1,2,...,n \end{aligned} \qquad (5\text{-}9)$$

$$\begin{aligned} X_{Qi} &= X_Q(i\Delta) \\ &= \theta s_Q(i\Delta) + W_Q(i\Delta) \\ &\triangleq \theta s_{Qi} + W_{Qi}, \quad i = 1,2,...,n \end{aligned} \qquad (5\text{-}10)$$

We find that the components X_{Ii} and X_{Qi} are individually described by the observation model of (2-1), since s_{Ii} and s_{Qi} are known quantities. Indeed, we will now assume that the in-phase noise components W_{Ii}, $i = 1,2,...,n$, form a sequence of i.i.d. random variables, governed by a common univariate density function f_L. Similarly, we will assume that the quadrature noise components are i.i.d. and governed by the same univariate noise density function f_L. The restriction that the sampling period Δ be such as to result in independent samples is necessary to enable us to obtain analytical results for non-Gaussian noise processes. If the noise bandwidth B is several times larger than the signal bandwidth B, such a sampling rate would be quite realistic to use.

From general results on representation of narrowband noise processes in the form of (5-2), we know that the in-phase and quadrature processes $W_I(t)$ and $W_Q(t)$ are *uncorrelated* at any time [see, for example, Van Trees, 1971]. For Gaussian input noise $W(t)$ this implies that W_{Ii} and W_{Qi} are *independent* Gaussian random variables. For non-Gaussian noise processes, however, independence of the in-phase and quadrature component samples at any time instant cannot be reasonably assumed. We will therefore assume here that W_{Ii} and W_{Qi} have, for each i, some common bivariate probability density function f_{IQ}. To complete our description of the statistics of the noise components, we will assume finally that the two-dimensional noise sample (W_{Ii}, W_{Qi}), $i = 1,2,...,n$, form a set of i.i.d. random vectors governed by the common probability density function f_{IQ}, for which the marginal densities are f_L.

Let $\mathbf{X}_I = (X_{I1}, X_{I2}, ..., X_{In})$ be the vector of in-phase observations, and let $\mathbf{X}_Q$ be the vector of quadrature observations. Then we may formally state our problem of testing $\theta = 0$ versus $\theta > 0$ in the observation model (5-9) and (5-10) as one of testing a null hypothesis H_2 versus an alternative hypothesis K_2 on the joint density function $f_{\mathbf{X}_I, \mathbf{X}_Q}$ of $\mathbf{X}_I$ and $\mathbf{X}_Q$:

$$H_2: f_{\mathbf{X}_I, \mathbf{X}_Q}(\mathbf{x}_I, \mathbf{x}_Q) = \prod_{i=1}^{n} f_{IQ}(x_{Ii}, x_{Qi}) \qquad (5\text{-}11)$$

$$K_2: f_{\mathbf{X}_I, \mathbf{X}_Q}(\mathbf{x}_I, \mathbf{x}_Q) = \prod_{i=1}^{n} f_{IQ}(x_{Ii} - \theta s_{Ii}, x_{Qi} - \theta s_{Qi}) \qquad (5\text{-}12)$$

for any $\theta > 0$ under K_1, with $\mathbf{s}_I$ and $\mathbf{s}_Q$ specified. Here $\mathbf{s}_I$ and $\mathbf{s}_Q$ are the vectors $(s_{I1}, s_{I2}, ..., s_{In})$ and $(s_{Q1}, s_{Q2}, ..., s_{Qn})$ of in-phase and quadrature signal components, respectively, which are samples of $s_I(t)$ and $s_Q(t)$. Notice the correspondence between the above pair of hypotheses and (H_1, K_1) of (2-13) and (2-14).

Let us make one interesting observation about the mechanism generating the data for this detection problem. Suppose there was no phase modulation in (5-1), so that $\phi(t) \equiv 0$. Then only the in-phase components X_{Ii} would be used, since the s_{Qi} are now all zero. The in-phase components can in this case be obtained by sampling the input process directly at the positive peak values of the carrier; one can use a sampling period very close to Δ and achieve this, since we have $\Delta \gg 2\pi/\omega_0$. In the more general case of phase modulation, this scheme can be modified by adding a sample exactly a quarter cycle after each of the "in-phase" samples. This procedure will generate bivariate observation components (X_{Ii}, X_{Qi}) which may also be approximated by the model given by (5-9) and (5-10), since the waveforms $s_I(t)$, $s_Q(t)$, $W_I(t)$, and $W_Q(t)$ are low-pass waveforms relative to the carrier frequency.

5.3 Locally Optimum Detection

From our statement of the generalized Neyman-Pearson lemma in Section 1.4 of Chapter 1 we know that under some general regularity conditions on f_{IQ} the LO detector uses as a test statistic the derivative at $\theta = 0$ of the log of the joint probability density function of $\mathbf{X}_I$ and $\mathbf{X}_Q$ under K_2. This was illustrated for known-signal detection in Section 2.3. In the present case we get the LO test statistic as

$$\lambda_{LO}(\mathbf{X}_I, \mathbf{X}_Q) = \frac{d}{d\theta} \left[\ln \prod_{i=1}^{n} f_{IQ}(X_{Ii} - \theta s_{Ii}, X_{Qi} - \theta s_{Qi}) \right]\Bigg|_{\theta = 0}$$

$$= \sum_{i=1}^{n} \frac{\frac{d}{d\theta}[f_{IQ}(X_{Ii} - \theta s_{Ii}, X_{Qi} - \theta s_{Qi})]\Big|_{\theta=0}}{f_{IQ}(X_{Ii}, X_{Qi})}$$

$$= \sum_{i=1}^{n} s_{Ii} \left[\frac{-\frac{\partial}{\partial X_{Ii}} f_{IQ}(X_{Ii}, X_{Qi})}{f_{IQ}(X_{Ii}, X_{Qi})} \right]$$

$$+ \sum_{i=1}^{n} s_{Qi} \left[\frac{-\frac{\partial}{\partial X_{Qi}} f_{IQ}(X_{Ii}, X_{Qi})}{f_{IQ}(X_{Ii}, X_{Qi})} \right] \quad (5\text{-}13)$$

Defining $g_{LO,I}$ and $g_{LO,Q}$ as the real-valued functions of bivariate inputs

$$g_{LO,I}(u,v) = \frac{-\frac{\partial}{\partial u} f_{IQ}(u,v)}{f_{IQ}(u,v)} \quad (5\text{-}14)$$

and

$$g_{LO,Q}(u,v) = \frac{-\frac{\partial}{\partial v} f_{IQ}(u,v)}{f_{IQ}(u,v)} \quad (5\text{-}15)$$

we get

$$\lambda_{LO}(\mathbf{X}_I, \mathbf{X}_Q) = \sum_{i=1}^{n} s_{Ii} \, g_{LO,I}(X_{Ii}, X_{Qi})$$

$$+ \sum_{i=1}^{n} s_{Qi} \, g_{LO,Q}(X_{Ii}, X_{Qi}) \quad (5\text{-}16)$$

It is instructive to compare $\lambda_{LO}(\mathbf{X}_I,\mathbf{X}_Q)$ of (5-16) for the narrowband signal case with $\lambda_{LO}(\mathbf{X})$ of (2-32). Note that because X_{Ii} and X_{Qi} are generally dependent, the nonlinear operations in the in-phase and quadrature signal channels each involve both in-phase and quadrature observation components. If we assume that W_{Ii} and W_{Qi} are independent, so that $f_{IQ}(u,v) = f_L(u)f_L(v)$, we get

$$g_{LO,I}(u,v) = \frac{-f_L'(u)}{f_L(u)} \qquad (5\text{-}17)$$

and

$$g_{LO,Q}(u,v) = \frac{-f_L'(v)}{f_L(v)} \qquad (5\text{-}18)$$

The processing is *decoupled* in the in-phase and quadrature signal channels, and $\lambda_{LO}(\mathbf{X}_i,\mathbf{X}_Q)$ is then essentially no different from $\lambda_{LO}(\mathbf{X})$ of (2-32). In particular, suppose that the quadrature component of the signal waveform is absent, that is $s_Q(t)$ is zero in (5-5). This implies that the known signal component in (5-1) is simply $v(t)\cos\omega_0 t$ or $s_I(t)\cos\omega_0 t$. Since the s_{Qi} are now zero we find that the decoupled test statistic (under the assumption that W_{Ii} and W_{Qi} are independent) now simplifies further to

$$\lambda_{LO}(\mathbf{X}_I,\mathbf{X}_Q) = \sum_{i=1}^n s_{Ii}\left[\frac{-f_L'(X_{Ii})}{f_L(X_{Ii})}\right] \qquad (5\text{-}19)$$

which is not a function of $\mathbf{X}_Q$. Thus in this particular case (no quadrature signal components *and* W_{Ii}, W_{Qi} independent) the resulting test statistic is exactly equivalent to $\lambda_{LO}(\mathbf{X})$ of (2-32) applied to the in-phase observation components.

We note that it would have been possible to obtain the result (5-16) using the general form of the multivariate LO detector test statistic developed in Section 2.6 of Chapter 2.

Circularly Symmetric Bivariate Noise Density

A more interesting model is obtained if we assume that f_{IQ} is a *circularly symmetric* bivariate density function, which means that $f_{IQ}(u,v)$ can be written as a function of $\sqrt{u^2 + v^2}$:

$$f_{IQ}(u,v) = h(r)\Big|_{r=\sqrt{u^2+v^2}} \qquad (5\text{-}20)$$

A familiar example of such a density function is the zero-mean bivariate Gaussian density function for uncorrelated random variables with a common variance,

$$f_{IQ}(u,v) = \frac{1}{2\pi\sigma^2} e^{-\frac{1}{2\sigma^2}(u^2+v^2)}$$

$$= \frac{1}{2\pi\sigma^2} e^{-r^2/2\sigma^2} \Big|_{r=\sqrt{u^2+v^2}} \quad (5\text{-}21)$$

Note that in general the function $f(r) = 2\pi r h(r)$, for $r \geq 0$, is a probability density function. In fact, it is the probability density function of the noise envelope $\sqrt{W_{Ii}^2 + W_{Qi}^2}$. This provides one method of generating different bivariate circularly symmetric density functions f_{IQ}.

With circular symmetry we find that $g_{LO,I}$ of (5-14) becomes

$$g_{LO,I}(u,v) = \frac{-\frac{\partial}{\partial u} h(\sqrt{u^2+v^2})}{h(\sqrt{u^2+v^2})}$$

$$= \frac{-u}{\sqrt{u^2+v^2}} \frac{h'(\sqrt{u^2+v^2})}{h(\sqrt{u^2+v^2})} \quad (5\text{-}22)$$

Now let R_i be the observation *envelope* at the i-th sampling time,

$$R_i = \sqrt{X_{Ii}^2 + X_{Qi}^2} \quad (5\text{-}23)$$

Then we can write $g_{LO,I}(X_{Ii}, X_{Qi})$ as

$$g_{LO,I}(X_{Ii}, X_{Qi}) = X_{Ii} \left[\frac{-h'(R_i)}{R_i h(R_i)} \right] \quad (5\text{-}24)$$

and in a similar way we get

$$g_{LO,Q}(X_{Ii}, X_{Qi}) = X_{Qi} \left[\frac{-h'(R_i)}{R_i h(R_i)} \right] \quad (5\text{-}25)$$

Defining the function $\tilde{g}_{LO}$ of a single variable to be

$$\tilde{g}_{LO}(r) = \frac{-h'(r)}{r h(r)} \quad (5\text{-}26)$$

we get for our LO test statistic for circularly symmetric f_{IQ} the result

$$\lambda_{LO}(\mathbf{X}_I, \mathbf{X}_Q) = \sum_{i=1}^n \tilde{g}_{LO}(R_i) \left[s_{Ii} X_{Ii} + s_{Qi} X_{Qi} \right] \quad (5\text{-}27)$$

For the circularly symmetric Gaussian bivariate density function for (W_{Ii}, W_{Qi}) we have $h(r)$ from (5-21), and this gives

$$\tilde{g}_{LO}(r) = \frac{1}{\sigma^2} \qquad (5\text{-}28)$$

so that $\lambda_{LO}(\mathbf{X}_I, \mathbf{X}_Q)$ becomes equivalent to

$$\lambda_{LO}(\mathbf{X}_I, \mathbf{X}_Q) = \sum_{i=1}^{n} \left[s_{Ii} X_{Ii} + s_{Qi} X_{Qi} \right] \qquad (5\text{-}29)$$

(where the constant $1/\sigma^2$ has been dropped). This corresponds to the use of a matched-filter processor for known narrowband signals, the optimum scheme for Gaussian noise. The more general result of (5-27) therefore implies that the linear matched filtering is performed, for non-Gaussian circularly symmetric noise densities, on in-phase and quadrature observation components which have been *modified* by multiplication with a nonlinear function $\tilde{g}_{LO}$ of the instantaneous envelope. Figure 5.1 illustrates the structure of the LO detector for H_2 versus K_2 for a circularly symmetric noise density function f_{IQ}.

Generalized Narrowband Correlator Detectors

Once again, we note the correspondence between the system of Figure 5.1 and the one based on $\lambda_{LO}(\mathbf{X})$ of (2-32), which is of the generalized correlator (GC) type shown in Figure 2.6. Figure 5.1 suggests that a useful class of detectors is generated by allowing arbitrary nonlinearities $\tilde{g}$ in place $\tilde{g}_{LO}$. We may call such detectors *generalized narrowband correlator* (GNC) detectors. In the special case where $\tilde{g}$ is a *constant*, as in the case of the optimum scheme for Gaussian noise, the resulting detector may be called the *linear narrowband correlator* (LNC) detector. Notice that $\tilde{g}(R_i)$ scales X_{Ii} and X_{Qi} prior to their use in the linear part of the GNC detector. We may interpret this scaling as the formation of products $[R_i \tilde{g}(R_i)][X_{Ii}/R_i]$ and $[R_i \tilde{g}(R_i)][X_{Qi}/R_i]$ of *normalized* in-phase and quadrature components and the quantity $R_i \tilde{g}(R_i)$. It is then clear that a *constant* value for $\tilde{g}$ implies *linear* weighting of the normalized in-phase and quadrature components.

The test statistic $T_{GNC}(\mathbf{X}_I, \mathbf{X}_Q)$ of a GNC detector may therefore be expressed as

$$T_{GNC}(\mathbf{X}_I, \mathbf{X}_Q) = \sum_{i=1}^{n} R_i \tilde{g}(R_i) \left[s_{Ii} \frac{X_{Ii}}{R_i} + s_{Qi} \frac{X_{Qi}}{R_i} \right] \qquad (5\text{-}30)$$

Another special case which is immediately suggested by this

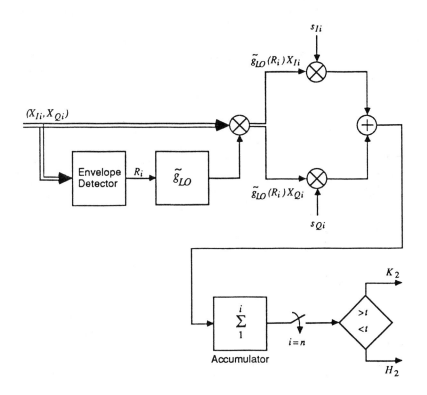

Figure 5.1 Locally Optimum Detector for Known Narrowband Signal in Narrowband Noise with a Circularly Symmetric Bivariate Noise Density Function

representation is that arising when $\tilde{g}(R_i) = 1/R_i$. The resulting test statistic leads to the *hard-limiter narrowband correlator* (HNC) detector, and is given by

$$T_{HNC}(\mathbf{X}_I, \mathbf{X}_Q) = \sum_{i=1}^{n} \left[s_{Ii} \frac{X_{Ii}}{R_i} + s_{Qi} \frac{X_{Qi}}{R_i} \right] \qquad (5\text{-}31)$$

The scaled components X_{Ii}/R_i and X_{Qi}/R_i represent a *unit-amplitude* vector. Thus the hard limiting here is effectively applied to the *envelope* of the input process and the resulting narrowband process converted to in-phase and quadrature components followed by sampling to produce X_{Ii}/R_i and X_{Qi}/R_i.

The HNC detector can be implemented by passing the input narrowband observation process through a hard limiter followed by a narrowband filter, after which its in-phase and quadrature components give the normalized sample values X_{Ii}/R_i and X_{Qi}/R_i. To see this, consider first any nonlinearity g applied directly to the narrowband process $X(t)$ followed by a narrowband filter, as shown in Figure 5.2. Let the input narrowband process be represented in terms of its envelope function $R(t)$ and phase $\Phi(t)$ as

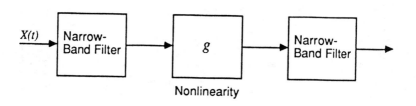

Figure 5.2 Narrowband System for Implementation of Envelope Nonlinearity

$$X(t) = R(t) \cos[\omega_0 t + \Phi(t)] \tag{5-32}$$

Now if we expand the function $g(r \cos \phi)$ in a Fourier series for the variable ϕ, we get

$$g(r \cos \phi) = a_0(r) + \sum_{k=1}^{\infty} a_k(r) \cos(k\phi) \tag{5-33}$$

where

$$a_0(r) = \frac{1}{2\pi} \int_{-\pi}^{\pi} g(r \cos \phi) \, d\phi \tag{5-34}$$

and

$$a_k(r) = \frac{1}{\pi} \int_{-\pi}^{\pi} g(r \cos \phi) \cos(k\phi) \, d\phi, \quad k = 1,2,... \tag{5-35}$$

From this we find that a narrowband filter with center frequency $f_0 = \omega_0/2\pi$ acting on $g\{R(t) \cos[\omega_0 t + \Phi(t)]\}$ produces the output

$$Y(t) = a_1[R(t)] \cos[\omega_0 t + \Phi(t)] \tag{5-36}$$

The in-phase and quadrature components of this waveform are, respectively,

$$Y_I(t) = a_1[R(t)] \frac{X_I(t)}{R(t)} \qquad (5\text{-}37)$$

and

$$Y_Q(t) = a_1[R(t)] \frac{X_Q(t)}{R(t)} \qquad (5\text{-}38)$$

Thus the HNC detector can be implemented using $g(x) = \text{sgn}(x)$, because in this case it follows easily that $a_1(r)$ is independent of r for $r > 0$.

The HNC detector is a particularly simple type of GNC detector and has been recognized as a useful detector when highly impulsive noise is present [e.g., Bello and Esposito, 1969, 1971]. Notice that it can be derived from (5-26) as the optimum detector for h given by

$$h(r) = \frac{1}{2\pi c^2} e^{-r/c}, \quad r \geq 0 \qquad (5\text{-}39)$$

for any $c > 0$. The HNC detector may be characterized as a *nonparametric* detector with a fixed false-alarm probability for a useful class of noise pdf's [Kassam, 1987].

Generalized Rayleigh Noise Envelope

We have noted that $2\pi r h(r)$ is the probability density function of the envelope of the noise, and that Gaussian noise generates the *Rayleigh* envelope density with h given by (5-21). We have also seen that the function h of (5-39) leads to an interesting special case, the HNC detector. These observations, together with our model for generalized Gaussian noise of Chapter 3, suggest consideration of a generalized Rayleigh envelope density for which the function h is now

$$h_k(r) = C(k) e^{-|r/B(k)|^k} \qquad (5\text{-}40)$$

where $k > 0$ controls the rate of decay of the function h_k and $B(k)$, $C(k)$ are chosen to make $2\pi r h_k(r)$ a valid envelope density function. It can be easily verified that for the corresponding univariate density functions f_L to have variance σ^2, we need

$$B(k) = \left[2\sigma^2 \frac{\Gamma(2/k)}{\Gamma(4/k)} \right]^{1/2} \qquad (5\text{-}41)$$

and

$$C(k) = \frac{k}{2\pi B^2(k)\Gamma(2/k)} \qquad (5\text{-}42)$$

Notice that $k = 2$ gives us the Rayleigh case, and for $k = 1$ we get

$$h_k(r) = \frac{3}{2\pi\sigma^2} e^{-\sqrt{3}r/\sigma} \tag{5-43}$$

which is the same as h of (5-39) with $c = \sigma/\sqrt{3}$.

From (5-26) we find that the LO nonlinearity for the generalized Rayleigh envelope density is

$$\tilde{g}_{LO}(r) = \frac{k}{\sigma^2} \left[\frac{\Gamma(4/k)}{2\Gamma(2/k)} \right]^{k/2} \left[\frac{r}{\sigma} \right]^{k-2} \tag{5-44}$$

The normalized characteristic $\sigma^2 \tilde{g}_{LO}(r)$ is plotted as a function of r/σ for different k in Figure 5.3. From this figure it is obvious that, as we would have expected, for "heavy-tailed" envelope densities a more severe limiting of the envelope should be employed for optimum performance.

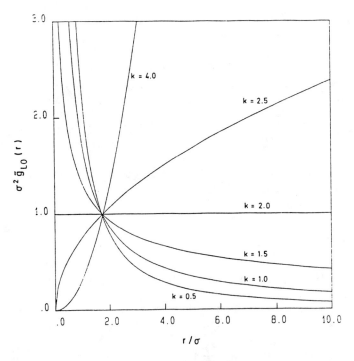

Figure 5.3 Normalized Locally Optimum Nonlinearities for Generalized Rayleigh Noise Envelope

As in the case of the HNC detector, the GNC detector requiring some particular envelope nonlinearity $\tilde{g}$ may be implemented as a nonlinearity g acting directly on the input process (followed by narrowband filtering and matched filtering). The function g has to have its Fourier series component function a_1 satisfy the requirement $r\tilde{g}(r) = a_1(r)$. Alternatively, the GNC detector can always be implemented by first obtaining the in-phase and quadrature input components and using the scheme of Figure 5.1.

5.4 Asymptotic Performance Analysis

We will now obtain a general expression for the efficacy of a GNC detector for a circularly symmetric bivariate noise density function f_{IQ}. This will allow us to compare the performance of a GNC detector with the LNC detector on the basis of the ARE.

Let us define random variables Z_i as

$$Z_i = \tilde{g}(R_i)(s_{Ii} X_{Ii} + s_{Qi} X_{Qi}) \tag{5-45}$$

To find the efficacy of a GNC detector we need to find the expected value of Z_i under the alternative hypothesis K_2 and its variance under the null hypothesis H_2. We are actually concerned with subsets of the density functions allowed under H_2 and K_2, since we will be restricting attention to the class of circularly symmetric bivariate densities f_{IQ} to simplify the analysis. Now it is clear that because of circular symmetry the expected value of Z_i under the null hypothesis is zero. To find its variance we therefore need $E\{Z_i^2 \mid \theta = 0\}$. For this we have

$$\begin{aligned}E\{Z_i^2 \mid \theta = 0\} &= E\{\tilde{g}^2(R_i)(s_{Ii} X_{Ii} + s_{Qi} X_{Qi})^2 \mid \theta = 0\} \\ &= E\left\{\tilde{g}^2(R_i) R_i^2 \left(s_{Ii}^2 \frac{X_{Ii}^2}{R_i^2} + s_{Qi}^2 \frac{X_{Qi}^2}{R_i^2}\right) \Big| \theta = 0\right\}\end{aligned} \tag{5-46}$$

This follows because $\tilde{g}^2(R_i) X_{Ii} X_{Qi}$ once again has mean value zero for $\theta = 0$ under the circular symmetry assumption. The circular symmetry for f_{IQ} also implies that the envelope R_i and phase $\Phi_i = \tan^{-1}(X_{Ii}/X_{Qi})$ are *independent* random variables when $\theta = 0$, the phase being uniformly distributed on $[0, 2\pi]$ in this case. This gives

$$V\{Z_i \mid \theta = 0\} = E\{R_i^2 \tilde{g}^2(R_i) \mid \theta = 0\} E\{s_{Ii}^2 \sin^2\Phi_i + s_{Qi}^2 \cos^2\Phi_i \mid \theta = 0\}$$

$$= \frac{1}{2}(s_{Ii}^2 + s_{Qi}^2) \int_0^\infty r^2 \tilde{g}^2(r) 2\pi r h(r)\, dr$$

$$= \pi(s_{Ii}^2 + s_{Qi}^2) \int_0^\infty r^3 \tilde{g}^2(r) h(r) \, dr \qquad (5\text{-}47)$$

where $2\pi r h(r)$ is, from (5-20), the probability density function of R_i under the null hypothesis.

For the mean value under the alternative hypothesis we get

$$E\{Z_i\} = \int_{-\infty}^{\infty}\int_{-\infty}^{\infty} \tilde{g}(\sqrt{u^2+v^2})(s_{Ii} u + s_{Qi} v) f_{IQ}(u - \theta s_{Ii}, v - \theta s_{Qi}) \, du \, dv \qquad (5\text{-}48)$$

Now we have

$$\frac{d}{d\theta} f_{IQ}(u - \theta s_{Ii}, v - \theta s_{Qi})\bigg|_{\theta=0}$$
$$= s_{Ii} g_{LO,I}(u,v) f_{IQ}(u,v) + s_{Qi} g_{LO,Q}(u,v) f_{IQ}(u,v)$$
$$= \frac{(-s_{Ii} u - s_{Qi} v) h'(\sqrt{u^2+v^2})}{\sqrt{u^2+v^2}} \qquad (5\text{-}49)$$

so that

$$\frac{d}{d\theta} E\{Z_i\}\bigg|_{\theta=0}$$
$$= -\int_{-\infty}^{\infty}\int_{-\infty}^{\infty} \tilde{g}(\sqrt{u^2+v^2}) \frac{(s_{Ii} u + s_{Qi} v)^2}{\sqrt{u^2+v^2}} h'(\sqrt{u^2+v^2}) \, du \, dv$$
$$= -\int_0^{2\pi}\int_0^{\infty} \tilde{g}(r) h'(r)(s_{Ii}^2 \sin^2\phi + s_{Ii} s_{Qi} \sin 2\phi + s_{Qi}^2 \cos^2\phi) r^2 \, dr \, d\phi$$
$$= -\pi(s_{Ii}^2 + s_{Qi}^2)\int_0^{\infty} r^2 \tilde{g}(r) h'(r) \, dr \qquad (5\text{-}50)$$

Using (5-47) and (5-50) in the general expression (1-41), we get for the efficacy ξ_{GNC} of the GNC detector for circularly symmetric f_{IQ} the result

$$\xi_{GNC} = \pi \frac{\left[\int_0^{\infty} r^2 \tilde{g}(r) h'(r) \, dr\right]^2}{\int_0^{\infty} r^3 \tilde{g}^2(r) h(r) \, dr} \tilde{P}_*^2 \qquad (5\text{-}51)$$

where $\tilde{P}_*^2$ is a measure of the average signal power,

$$\tilde{P}_s^2 = \lim_{n \to \infty} \frac{1}{n} \sum_{i=1}^{n} (s_{Ii}^2 + s_{Qi}^2) \qquad (5\text{-}52)$$

As in our earlier result for the GC detectors, we define the normalized efficacy $\tilde{\xi}_{GNC}$ by

$$\begin{aligned}\tilde{\xi}_{GNC} &= \xi_{GNC}/\tilde{P}_s^2 \\ &= \pi \frac{\left[\int_0^\infty r^2 \tilde{g}(r) h'(r)\, dr\right]^2}{\int_0^\infty r^3 \tilde{g}^2(r) h(r)\, dr}\end{aligned} \qquad (5\text{-}53)$$

If we reduce $\tilde{P}_s^2$ in (5-52) by a factor of 2 for this narrowband signal situation, the expression for the normalized efficacy $\tilde{\xi}_{GNC}$ above would have 2π in it in place of π. Notice that the efficacy $\tilde{\xi}_{GNC}$ depends on the nonlinearity $\tilde{g}$ characterizing the GNC detector, and on the function h defining the bivariate density function f_{IQ} of the in-phase and quadrature noise components through (5-20). For the special case where $\tilde{g}$ is the LO nonlinearity $\tilde{g}_{LO}$, the efficacy $\tilde{\xi}_{GNC}$ becomes

$$\tilde{\xi}_{GNC,LO} = \pi \int_0^\infty \left[\frac{h'(r)}{h(r)}\right]^2 h(r) r\, dr \qquad (5\text{-}54)$$

Note that the Schwarz inequality gives

$$\begin{aligned}&\left[\int_0^\infty r \tilde{g}(r) \frac{h'(r)}{h(r)} h(r) r\, dr\right]^2 \\ &\leq \int_0^\infty r^2 \tilde{g}^2(r) h(r) r\, dr \int_0^\infty \left[\frac{h'(r)}{h(r)}\right]^2 h(r) r\, dr\end{aligned} \qquad (5\text{-}55)$$

with equality for $r\tilde{g}(r) = Kh'(r)/h(r)$ for any constant $K \neq 0$; from (5-55) it follows that

$$\tilde{\xi}_{GNC} \leq \tilde{\xi}_{GNC,LO} \qquad (5\text{-}56)$$

The nonlinearity $\tilde{g}_{LO}$ for $\tilde{g}$ in the GNC detector makes it the *asymptotically optimum* detector for this signal detection problem.

Letting $\tilde{g}$ be any *constant* in (5-53) we get the normalized efficacy $\tilde{\xi}_{LNC}$ of the LNC detector as

$$\tilde{\xi}_{LNC} = \pi \frac{\left[\int_0^\infty r^2 h'(r)\, dr\right]^2}{\int_0^\infty r^3 h(r)\, dr} \qquad (5\text{-}57)$$

Now we have

$$\int_0^\infty r^3 h(r)\, dr = \frac{1}{2\pi} \int_{-\infty}^\infty \int_{-\infty}^\infty (u^2 + v^2) f_{IQ}(u,v)\, du\, dv$$

$$= \frac{1}{\pi} \int_{-\infty}^\infty u^2 f_L(u)\, du$$

$$= \frac{\sigma^2}{\pi} \qquad (5\text{-}58)$$

where σ^2 is the variance of the in-phase and quadrature noise components. In addition, we may also interpret $2\sigma^2$ as the *second moment* of the noise envelope, since $2\pi r h(r)$ is the noise envelope density. Then, assuming that $r^2 h(r)$ is zero in the limits $r \to 0$ and $r \to \infty$, we get from (5-57) the result

$$\tilde{\xi}_{LNC} = \frac{1}{\sigma^2} \qquad (5\text{-}59)$$

Again we find a correspondence between this result and that of (3-6) for the LC detector efficacy.

As we have mentioned before, another special case which is of practical significance is that of the *hard-limiter narrowband correlator* (HNC) detector, obtained with $\tilde{g}(r) = 1/r$. For this we easily get the normalized efficacy to be

$$\tilde{\xi}_{HNC} = 2\pi^2 \left[\int_0^\infty h(r)\, dr\right]^2 \qquad (5\text{-}60)$$

From these results for the efficacies we can easily obtain numerical comparisons of the asymptotic performance of different pairs of GNC detectors under different noise conditions, using the ARE. We will now look at the particular class of generalized Rayleigh noise envelope densities and obtain some specific results.

Generalized Rayleigh Noise Envelope

The efficacy $\tilde{\xi}_{GNC,LO}$ of the LO detector operating under the noise condition for which it is optimum can be obtained from (5-54). For generalized Rayleigh noise envelope we get

$$\tilde{\xi}_{GNC,LO} = \pi \int_0^\infty \frac{k^2}{\sigma^{2k}} \left[\frac{\Gamma(4/k)}{2\Gamma(2/k)} \right]^k r^{2k-1} h_k(r)\, dr$$

$$= \frac{k^2}{2B^2(k)\Gamma(2/k)} \quad (5\text{-}61)$$

For $k = 2$ this gives the result $1/\sigma^2$, which is therefore the efficacy of the LNC detector for Gaussian noise. From (5-59) we know that $\tilde{\xi}_{LNC} = 1/\sigma^2$ for any circularly symmetric density function. Thus the ARE of the LO GNC detector for h_k relative to the LNC detector, when h_k describes the noise envelope density, is

$$ARE_{LO,LNC} = \frac{k^2 \Gamma(4/k)}{4\Gamma^2(2/k)} \quad (5\text{-}62)$$

The minimum value of this ARE as a function of k is unity at $k = 2$, and its value for $k = 1$ is 1.5. It increases rapidly for further reduction in the value of k; for example, $k = 2/3$ makes the $ARE_{LO,LNC} \approx 3.3$ and $k = 0.5$ makes it 8.75. Thus a very considerable improvement in performance can be obtained by using the LO GNC detector rather than the LNC detector in highly impulsive noise.

From (5-60) we find that for generalized Rayleigh noise envelopes the HNC detector has efficacy

$$\tilde{\xi}_{HNC} = 2\pi^2 \left[\int_0^\infty h_k(r)\, dr \right]^2$$

$$= \frac{\Gamma(4/k)\Gamma^2(1/k)}{4\sigma^2 \Gamma^3(2/k)} \quad (5\text{-}63)$$

so that we get

$$ARE_{HNC,LNC} = \frac{\Gamma(4/k)\Gamma^2(1/k)}{4\Gamma^3(2/k)} \quad (5\text{-}64)$$

Note that in terms of $ARE_{LO,LNC}$ of (5-62), we have

$$ARE_{HNC,LNC} = ARE_{LO,LNC} \frac{\Gamma^2(1/k)}{k^2 \Gamma(2/k)} \quad (5\text{-}65)$$

Now the term $\Gamma^2(1/k)/[k^2\Gamma^2(2/k)]$ is fairly close to unity for $2/3 \leq k \leq 2$. For example, its value at $k = 1$ is unity, for $k = 2$ its value is $\pi/4$ and for $k = 2/3$ its value is $(9/8)(\pi/4)$. This indicates that the HNC detector is a very good alternative to the LNC detector for operation in noise which may have an impulsive component present.

The generalized Rayleigh noise envelope density model does not allow one to explicitly control the *univariate* density function f_L of the in-phase and quadrature noise terms. If this is desired, another model can be used, as described by Modestino and Ningo [1979]. In the alternative model the *characteristic function* of a generalized univariate noise density leads to the function h for the envelope density. It is also possible to consider a *mixture* noise-envelope density model, although we shall not go into further details at this point. The class of generalized Rayleigh envelope densities illustrates quite clearly the advantage obtainable by using a GNC detector or even the simple HNC detector for known narrowband signals in non-Gaussian noise.

In a recent paper Fedele, Izzo, and Paura [1984] consider locally optimum diversity detection of a narrowband signal observed at multiple locations, with different random amplitudes. They also consider a suboptimum detector which uses the separate decisions made by individual locally optimum detectors, and give numerical performance results. Also of interest are recent papers by Enge [1985, 1986] on narrowband signal detection in correlated and circularly symmetric noise.

5.5 Asymptotically Optimum Envelope Quantization

Under the assumption of circular symmetry of the bivariate density function of the in-phase and quadrature noise components, we have found that the locally optimum detector for a known narrowband signal requires a nonlinear version of the observation envelope samples. If we express the LO test statistic $\lambda_{LO}(\mathbf{X}_I,\mathbf{X}_Q)$ of (5-27) in the form of the test statistic $T_{GNC}(\mathbf{X}_I,\mathbf{X}_Q)$ of (5-30), we find that the nonlinearity is $r\tilde{g}_{LO}(r)$, $r \geq 0$. For the Rayleigh envelope density this is a linear function of r, whereas it is a constant for the case where h in the envelope density function is an exponential function.

Both for simplified implementation and to obtain a detection scheme which has characteristics in between those of the LNC detector and those of the HNC detector, we can use a piecewise-constant quantizer characteristic $\tilde{q}$ to obtain specifically the *quantizer narrowband correlator* (QNC) detector test statistic

$$T_{QNC}(\mathbf{X}_I,\mathbf{X}_Q) = \sum_{i=1}^{n} \tilde{q}(R_i)\left[s_{Ii}\frac{X_{Ii}}{R_i} + s_{Qi}\frac{X_{Qi}}{R_i}\right] \quad (5\text{-}66)$$

This is, of course, a particular case of the GNC detector, with $R_i\,\tilde{g}(R_i)$ in (5-30) replaced by $\tilde{q}(R_i)$. For M-interval envelope quantization we can define $\tilde{q}$ in terms of output levels

$l_1, l_2, \ldots, l_M$ and input breakpoints $t_0, t_1, \ldots, t_M$ with $t_0 \triangleq 0$, $t_M \triangleq \infty$ and $t_j \leq t_{j+1}$. The value of $\tilde{q}(r)$ is l_j for $t_{j-1} < r \leq t_j$.

Now from (5-53), the normalized efficacy $\tilde{\xi}_{QNC}$ of the QNC detector is

$$\tilde{\xi}_{QNC} = \pi \frac{\left[\sum_{j=1}^{M} l_j \int_{t_{j-1}}^{t_j} rh'(r)\, dr\right]^2}{\sum_{j=1}^{M} l_j^2 \int_{t_{j-1}}^{t_j} rh(r)\, dr}$$

$$= \frac{1}{2} \frac{\left[\sum_{j=1}^{M} l_j \int_{t_{j-1}}^{t_j} r\tilde{g}_{LO}(r) f_E(r)\, dr\right]^2}{\sum_{j=1}^{M} l_j^2 \left[F_E(t_j) - F_E(t_{j-1})\right]} \qquad (5\text{-}67)$$

where $f_E(r) = 2\pi r h(r)$ is the noise envelope density function and F_E is the corresponding distribution. For a *given* set of breakpoints we can obtain the LO levels as we did in the low-pass signal case in Section 4.2 of Chapter 4. These are also the levels maximizing $\tilde{\xi}_{QNC}$. It is easy to show that a set of levels maximizing $\tilde{\xi}_{QNC}$ can be defined as

$$l_j = \frac{\int_{t_{j-1}}^{t_j} r\tilde{g}_{LO}(r) f_E(r)\, dr}{F_E(t_j) - F_E(t_{j-1})}, \quad j = 1, 2, \ldots, M \qquad (5\text{-}68)$$

and that the resulting maximum $\tilde{\xi}_{QNC}$ for given breakpoints is

$$\tilde{\xi}_{QNC,LO} = \frac{1}{2} \sum_{j=1}^{M} \frac{\left[\int_{t_{j-1}}^{t_j} r\tilde{g}_{LO}(r) f_E(r)\, dr\right]^2}{F_E(t_j) - F_E(t_{j-1})} \qquad (5\text{-}69)$$

It is interesting to compare these results on envelope quantization for the narrowband-signal case with those of Section 4.2 in Chapter 4 for the low-pass case. As we did there, we can further optimize the quantizer design and get an AO QNC detector by maximizing $\tilde{\xi}_{QNC,LO}$ with respect to the breakpoints t_j. It is possible to show that this AO quantization is also a "best-fit" quantization of the LO characteristic $r\tilde{g}_{LO}(r)$.

We have noted earlier that a GNC detector may be implemented by using an appropriate nonlinearity acting on the input waveform followed by narrowband filtering, after which a linear matched filter can be used. If the nonlinearity is a hard limiter we get the HNC detector and if we use no input nonlinearity we get the LNC detector. In general, however, a QNC detector is *not* obtained by using a quantizer as the input nonlinearity [Problem 5.2]. The use of an input quantizer does, however, also produce a useful detector structure with performance in between that of the HNC and LNC detectors.

In the next chapter, we will look into another input-data quantization problem giving results which again correspond very closely to the results we have discussed here and in Chapter 4.

5.6 Locally Optimum Bayes Detection

In this section we will give results for detection of known narrowband signals which are similar to those we discussed for low-pass signals based on the Bayes criterion in Section 2.5 of Chapter 2. To obtain more compact expressions for our test statistics, which will also aid us in interpreting the results, let us first introduce the definitions

$$\tilde{s}_i = s_{Ii} + js_{Qi} \tag{5-70}$$

$$\tilde{X}_i = X_{Ii} + jX_{Qi} \tag{5-71}$$

and

$$\tilde{W}_i = W_{Ii} + jW_{Qi} \tag{5-72}$$

Thus the $\tilde{}$ variables are complex quantities formed from the corresponding in-phase and quadrature components, of the signal, observation, and noise quantities defined in Section 5.2. Note that the observation model of (5-9) and (5-10) may now be written as

$$\tilde{X}_i = \theta \tilde{s}_i + \tilde{W}_i, \quad i = 1,2,...,n \tag{5-73}$$

We will find this complex-variable formulation useful in the next chapter also.

It is also convenient to define a complex-valued function (nonlinearity) incorporating $g_{LO,I}(u,v)$ and $g_{LO,Q}(u,v)$ of (5-14) and (5-15) as follows:

$$\tilde{g}_{LO,IQ}(u + jv) = g_{LO,I}(u,v) + jg_{LO,Q}(u,v) \tag{5-74}$$

This allows us to express the general LO statistic of (5-16) in the form

$$\lambda_{LO}(\tilde{\mathbf{X}}) = \sum_{i=1}^{n} \text{Re}\{\tilde{s}_i \, \tilde{g}^*_{LO,IQ}(\tilde{X}_i)\} \qquad (5\text{-}75)$$

where the * denotes a complex-conjugate operation. In (5-75) the vector $\tilde{\mathbf{X}}$ is $(\tilde{X}_1, \tilde{X}_2, \ldots, \tilde{X}_n)$.

Observe that for a circularly symmetric bivariate noise density function f_{IQ} we have

$$\tilde{g}_{LO,IQ}(\tilde{X}_i) = \tilde{X}_i \, \tilde{g}_{LO}(R_i)$$
$$= \tilde{X}_i \, \tilde{g}_{LO}(|\tilde{X}_i|) \qquad (5\text{-}76)$$

where $\tilde{g}_{LO}$ was defined in (5-26). In this case the LO test statistic of (5-75) becomes

$$\lambda_{LO}(\tilde{\mathbf{X}}) = \sum_{i=1}^{n} \tilde{g}_{LO}(|\tilde{X}_i|) \, \text{Re}\{\tilde{s}_i \tilde{X}_i^*\} \qquad (5\text{-}77)$$

This is the same as $\lambda_{LO}(\mathbf{X}_I, \mathbf{X}_Q)$ of (5-27).

Let us now modify our observation model of (5-73), in which so far we have been assuming either that the $\tilde{s}_i$ are some specified set of signals and θ has a positive value or that they are all equal to zero (that is, $\theta = 0$). We now assume that the known-signal vector $\tilde{\mathbf{s}} = (\tilde{s}_1, \tilde{s}_2, \ldots, \tilde{s}_n)$ is either $\tilde{\mathbf{s}}_0$ or $\tilde{\mathbf{s}}_1$, and is present in $\tilde{\mathbf{X}}$ with amplitude θ. Let us also use the same assumptions and notation that we used for the costs and *a priori* probabilities in Section 2.5 of Chapter 2. Then we can deduce easily that for a finite sample size n and vanishing signal strength parameter θ, the Bayes detector which minimizes the expected value of the Bayes risk implements the test

$$\theta \sum_{i=1}^{n} \text{Re}\{(\tilde{s}_{1i} - \tilde{s}_{0i}) \, \tilde{g}^*_{LO,IQ}(\tilde{X}_i)\} > k \qquad (5\text{-}78)$$

where $\exp(k) = (c_{10}p_0/c_{01}p_1)$. For the case of equal costs, $c_{10} = c_{01}$, and equal *a priori* probabilities, $p_0 = p_1$, the threshold constant k is 0 and θ may be dropped from the left-hand side above. The resulting locally optimum Bayes detector then has a structure quite similar to that of the LO Neyman-Pearson detector, with threshold set equal to zero.

In considering the combined asymptotic situation $\theta \to 0$ and $n \to \infty$ we again focus on the case where $\theta = \gamma/\sqrt{n}$ for some finite positive constant γ. Then by expanding in a Taylor series the counterpart in this narrowband detection problem of the expression (2-85) of Chapter 2, and assuming sufficient regularity conditions on the bivariate probability density function f_{IQ}, we obtain a result similar to that of (2-90). Specifically, we find for $n \to \infty$ with $\theta\sqrt{n} = \gamma$ that an asymptotically optimum Bayes test implements

$$\theta \sum_{i=1}^{n} \mathrm{Re}\{(\tilde{s}_{1i} - \tilde{s}_{0i}) \, \tilde{g}'_{LO,IQ}(\tilde{X}_i)\} > k + [\text{bias term}] \quad (5\text{-}79)$$

Under the assumption of circular symmetry of $f_{IQ}(u,v)$, we get the asymptotically optimum Bayes test

$$\theta \sum_{i=1}^{n} \tilde{g}_{LO}(|\tilde{X}_i|) \, \mathrm{Re}\{(\tilde{s}_{1i} - \tilde{s}_{0i}) \, \tilde{X}_i^*\}$$
$$> k + \frac{\gamma^2}{2}(\tilde{P}_1^2 - \tilde{P}_0^2) \tilde{I}(h) \quad (5\text{-}80)$$

where

$$\tilde{P}_j^2 = \lim_{n \to \infty} \frac{1}{n} \sum_{i=1}^{n} |\tilde{s}_{ji}|^2, \quad j = 0,1 \quad (5\text{-}81)$$

and

$$\tilde{I}(h) = \pi \int_0^\infty \left[\frac{h'(r)}{h(r)}\right]^2 h(r) r \, dr \quad (5\text{-}82)$$

[which is the same quantity as $\tilde{\xi}_{GNC,LO}$ of (5-54)].

The result of (5-80) is of the same form as that of (2-90). Once again we find that for $\tilde{P}_1^2 = \tilde{P}_0^2$, which is true for antipodal signaling, the LO Bayes detector is asymptotically optimum when $\theta = \gamma/\sqrt{n}$ and $n \to \infty$. More generally, (5-78) and (5-80) are not the same and the Bayes detector which is optimum for weak signals for some finite n requires an additional term in its threshold for it to be an asymptotically optimum detector.

PROBLEMS

Problem 5.1

Consider the HNC detector based on the use of $T_{HNC}(\mathbf{X}_I, \mathbf{X}_Q)$ of (5-31) for testing H_2 versus K_2 of (5-11) and (5-12), when f_{IQ} is circularly symmetric. Show that for any desired false-alarm probability α, a threshold can be set so that the HNC detector has this false-alarm probability for *any* null hypothesis noise-only circularly symmetric pdf.

Problem 5.2

It has been shown that with $g(x) = \text{sgn}(x)$ in Figure 5.2, the system corresponds to an envelope hard limiter. Now let $g(x)$ be a symmetric four-level quantizer function. What effect does the system of Figure 5.2 now have on the envelope of the input process?

Problem 5.3

Derive the result (5-54) for the normalized efficacy of an LO GNC detector for circularly symmetric noise pdf using the results of Section 2.6 in Chapter 2.

Problem 5.4

Let f_{IQ} be a circularly symmetric noise pdf for which the corresponding envelope pdf is the mixture

$$f_E(r) = 2\pi r \left[(1 - \epsilon) h_2(r;1) + \epsilon h_2(r;\sigma^2)\right]$$

where $h_2(r;\sigma^2)$ is the pdf of (5-40) with $k = 2$. Find the ARE of the HNC detector relative to the LNC detector in testing H_2 versus K_2 for this noise pdf. Sketch the result as a function of σ^2 for fixed ϵ.

Sketch also the LO nonlinearity $\tilde{g}_{LO}$ for this mixture pdf, for $\epsilon = 0.1$ and $\sigma^2 = 10$.

Problem 5.5

Develop results for the optimum breakpoints in AO quantization of the envelope in the QNC detector of Section 5.5 corresponding to those for the AO QC detector for low-pass known signals of Section 4.2. Show that the AO envelope quantizer results in a "best-fit" quantization, in the sense of minimizing mean-squared error under the noise-only hypothesis, of the LO envelope nonlinearity $R_i \tilde{g}_{LO}(R_i)$.

Problem 5.6

The HNC detector is obtained when $\tilde{q}$ in (5-66) is a one-level quantizer, and its test statistic may be written as

$$T_{HNC}(\mathbf{X}_I, \mathbf{X}_Q) = \sum_{i=1}^{n} \text{Re}\{\tilde{s}_i \; \tilde{U}_i^*\}$$

where $\tilde{U}_i = \tilde{X}_i / R_i$ and $\tilde{s}_i, \tilde{x}_i$ are defined in (5-70) and (5-71). The $\tilde{U}_i$ are unit-amplitude vectors. Consider now a further uniform quantization of the $\tilde{U}_i$ into M values equally spaced on the unit circle (this corresponds to phase quantization). Obtain the efficacy of the resulting detector in testing H_2 versus K_2.

Chapter 6

DETECTION OF NARROWBAND SIGNALS WITH RANDOM PHASE ANGLES

6.1 Introduction

Although we considered the detection of a narrowband signal in narrowband noise in the last chapter, the assumption there remained that the signal was completely known. This meant, in particular, that the absolute carrier phase was precisely known for the narrowband signal. Now in many applications involving narrowband signals it may be much more realistic if the assumption of precise knowledge of the absolute carrier phase were dropped. In many such cases, moreover, it is reasonable to assume that this absolute carrier phase is a random variable uniformly distributed on $[0, 2\pi]$. Such a narrowband signal is called an *incoherent* narrowband signal, which is therefore a deterministic random process with one random parameter. We shall derive the LO detector for an incoherent narrowband signal in narrowband noise based on samples of the in-phase and quadrature observation components in Section 6.2. While the derivation of the LO detector is somewhat long, the detector test statistic is easily interpreted as an intuitively reasonable quantity.

In Section 6.3 a slightly different problem is considered, which may be described as the detection of *noncoherent narrowband pulses* based on the outputs of a filter matched to the pulse shape. This is a problem which arises, for instance, in radar detection of reflecting targets, when successive narrowband pulses have absolute phase angles which are *independent* uniformly distributed random variables. We conclude this chapter with an analysis in Section 6.4 of optimum data quantization for this important detection problem.

We shall see as we continue, in Chapter 7, that the results of this chapter form a bridge between those in the earlier chapters and those in the next chapter.

6.2 Detection of an Incoherent Signal

The narrowband signal that we will now consider is a random process with one random parameter, its phase. Although we can allow somewhat more generality and introduce a random amplitude as well, the resulting detector structure for locally optimum detection is essentially the same, as we will note later in this section.

6.2.1 The Observation Model

Generalizing our description of (5-1) to allow the narrowband signal to have a random absolute phase angle ψ, our continuous-time observation process becomes

$$X(t) = \theta v(t) \cos[\omega_0 t + \phi(t) + \psi] + W(t)$$
$$= X_I(t) \cos \omega_0 t + X_Q(t) \sin \omega_0 t \qquad (6\text{-}1)$$

where the in-phase and quadrature components are now

$$X_I(t) = \theta[s_I(t) \cos \psi + s_Q(t) \sin \psi] + W_I(t) \qquad (6\text{-}2)$$

$$X_Q(t) = \theta[-s_I(t) \sin \psi + s_Q(t) \cos \psi] + W_Q(t) \qquad (6\text{-}3)$$

The waveforms $s_I(t)$ and $s_Q(t)$ are defined by (5-4) and (5-5), and $W_I(t)$ and $W_Q(t)$ are the in-phase and quadrature components of the noise process $W(t)$. The sampled in-phase and quadrature observation components now become

$$X_{Ii} = \theta[s_{Ii} \cos \psi + s_{Qi} \sin \psi] + W_{Ii} \qquad (6\text{-}4)$$

and

$$X_{Qi} = \theta[-s_{Ii} \sin \psi + s_{Qi} \cos \psi] + W_{Qi} \qquad (6\text{-}5)$$

Here the s_{Ii} and s_{Qi} are, as before, the sampled values of $s_I(t)$ and $s_Q(t)$. Our assumption will be that the phase angle ψ is *uniformly* distributed on $[0, 2\pi]$, leading to the *incoherent* narrowband signal detection problem. We will not consider the partially coherent case of a non-uniform distribution for ψ. Our detection problem is now that of determining if θ is not equal to zero. Note that here we have a two-sided alternative ($\theta \neq 0$) to the null hypothesis $\theta = 0$. This arises because ψ is a random variable uniformly distributed on $[0, 2\pi]$, so that positive and negative values of θ cannot be distinguished using the observations X_{Ii} and X_{Qi}.

We define the row vectors **a**, **b** and $\mathbf{s}_i$ as

$$\mathbf{a} = (\cos \psi, \sin \psi) \qquad (6\text{-}6)$$

$$\mathbf{b} = (-\sin\psi, \cos\psi) \qquad (6\text{-}7)$$

and

$$\mathbf{s}_i = (s_{Ii}, s_{Qi}) \qquad (6\text{-}8)$$

and let $f(\theta;j)$ be, for given (X_{Ij}, X_{Qj}), the function of θ

$$f(\theta;j) = f_{IQ}(X_{Ij} - \theta s_j \mathbf{a}^T, X_{Qj} - \theta s_j \mathbf{b}^T) \qquad (6\text{-}9)$$

We will assume, as in the previous chapter, that f_{IQ} is a *circularly symmetric* bivariate probability density function for the in-phase and quadrature noise components. The joint probability density function of our $2n$ sampled in-phase and quadrature observation is

$$f_{\mathbf{X}_I,\mathbf{X}_Q}(\mathbf{X}_I,\mathbf{X}_Q) = E_\psi\left\{\prod_{i=1}^n f(\theta;i)\right\} \qquad (6\text{-}10)$$

under the assumption, again as in Chapter 5, that the samples (X_{Ii}, X_{Qi}) of the bivariate observation process form a sequence of independent random vectors for given ψ. In (6-10) the expectation is taken with respect to the random absolute phase ψ.

6.2.2 The Locally Optimum Detector

Before proceeding with the derivation of the LO detector test statistic it is convenient to define some further quantities and to introduce notation which will help simplify our exposition.

We define the functions f_{IQ}^i, f_{IQ}^{ii}, and f_{IQ}^{iq} by

$$f_{IQ}^i(u,v) = \frac{\partial}{\partial u} f_{IQ}(u,v) \qquad (6\text{-}11)$$

$$f_{IQ}^{ii}(u,v) = \frac{\partial^2}{\partial u^2} f_{IQ}(u,v) \qquad (6\text{-}12)$$

and

$$f_{IQ}^{iq}(u,v) = \frac{\partial^2}{\partial u \, \partial v} f_{IQ}(u,v) \qquad (6\text{-}13)$$

with similar definitions for f_{IQ}^q and f_{IQ}^{qq}. In addition, we will use the notation

$$f_{IQ}^i(j) = \frac{\partial}{\partial X_{Ij}} f_{IQ}(X_{Ij}, X_{Qj}) \qquad (6\text{-}14)$$

with a similar meaning for $f_{IQ}^{ii}(j)$, $f_{IQ}^{iq}(j)$, $f_{IQ}^{qi}(j)$, and $f_{IQ}^{qq}(j)$. With this notation we have $f(0;j) = f_{IQ}(j)$.

For the case of uniformly distributed ψ we find that **a** and **b** are zero-mean vectors, with

$$E\{\mathbf{a}^T \mathbf{a}\} = E\{\mathbf{b}^T \mathbf{b}\}$$

and

$$= \frac{1}{2}\begin{bmatrix} 1 & 0 \\ 0 & 1 \end{bmatrix} \tag{6-15}$$

$$E\{\mathbf{a}^T \mathbf{b}\} = [E\{\mathbf{b}^T \mathbf{a}\}]^T$$

$$= \frac{1}{2}\begin{bmatrix} 0 & 1 \\ -1 & 0 \end{bmatrix} \tag{6-16}$$

To complete our preliminaries, let us also note that since we will continue to assume that f_{IQ} is *circularly symmetric* we have, from (5-14), (5-24), and (5-26), that

$$\frac{f_{IQ}^i(j)}{f_{IQ}(j)} = -X_{Ij}\,\tilde{g}_{LO}(R_j) \tag{6-17}$$

and similarly

$$\frac{f_{IQ}^q(j)}{f_{IQ}(j)} = -X_{Qj}\,\tilde{g}_{LO}(R_j) \tag{6-18}$$

where R_j is the envelope value $\sqrt{X_{Ij}^2 + X_{Qj}^2}$. Furthermore, we obtain, since $f_{IQ}(u,v) = h(\sqrt{u^2 + v^2})$,

$$\begin{aligned} f_{IQ}^{ii}(u,v) &= \frac{\partial^2}{\partial u^2} h(\sqrt{u^2 + v^2}) \\ &= \frac{\partial}{\partial u}\left\{ h'(\sqrt{u^2 + v^2}) \frac{u}{\sqrt{u^2 + v^2}} \right\} \\ &= \frac{u^2 h''(\sqrt{u^2 + v^2})}{u^2 + v^2} + \frac{h'(\sqrt{u^2 + v^2})}{\sqrt{u^2 + v^2}} \\ &\quad - \frac{u^2 h'(\sqrt{u^2 + v^2})}{(u^2 + v^2)^{3/2}} \end{aligned} \tag{6-19}$$

so that

$$f_{I\ddot{Q}}(j) = X_{Ij}^2 \frac{h''(R_j)}{R_j^2} + \frac{h'(R_j)}{R_j} - X_{Ij}^2 \frac{h'(R_j)}{R_j^3} \qquad (6\text{-}20)$$

and similarly

$$f_{I\ddot{Q}}(j) = X_{Qj}^2 \frac{h''(R_j)}{R_j^2} + \frac{h'(R_j)}{R_j} - X_{Qj}^2 \frac{h'(R_j)}{R_j^3} \qquad (6\text{-}21)$$

Thus

$$\frac{\nabla^2 f_{IQ}(X_{Ij}, X_{Qj})}{f_{IQ}(X_{Ij}, X_{Qj})} = \frac{f_{I\ddot{Q}}(j) + f_{I\ddot{Q}}(j)}{f_{IQ}(j)}$$

$$= \frac{h''(R_j)}{h(R_j)} + \frac{h'(R_j)}{R_j h(R_j)} \qquad (6\text{-}22)$$

Our goal is to find the locally optimum detection test statistic for $\theta = 0$ versus $\theta \neq 0$. Generalizing our discussion of LO detectors in Section 2.3. of Chapter 2 let us obtain the first derivative of $f_{X_I, X_Q}(X_I, X_Q)$ with respect to θ, evaluated at $\theta = 0$. We get

$$\frac{d}{d\theta} f_{X_I, X_Q}(X_I, X_Q) = E_\psi \left\{ \sum_{i=1}^n f'(\theta; i) \prod_{j \neq i}^n f(\theta; j) \right\} \quad (6\text{-}23)$$

and find that this becomes zero at $\theta = 0$, since

$$E_\psi \{f'(0;i)\} = E_\psi \{-s_i \mathbf{a}^T f_{I\dot{Q}}(i) - s_i \mathbf{b}^T f_{I\dot{Q}}(i)\}$$
$$= 0 , \qquad (6\text{-}24)$$

a and **b** being zero-mean vectors.

This implies that the first derivative of the power function at $\theta = 0$ is always zero. In this case local optimality is interpreted as a condition leading to the maximum value of the *second* derivative of the power function at $\theta = 0$. The alternative interpretation is that we are expanding the likelihood ratio in powers of θ and retaining only the most significant term in θ. The second derivative of the joint density function is

$$\frac{d^2}{d\theta^2} f_{X_I,X_Q}(\mathbf{X}_I,\mathbf{X}_Q) = E_\psi \Bigg\{ \sum_{i=1}^{n} \Bigg[f''(\theta;i) \prod_{j \neq i}^{n} f(\theta;j)$$

$$+ f'(\theta;i) \sum_{j \neq i}^{n} f'(\theta;j) \prod_{\substack{k \neq j \\ k \neq i}}^{n} f(\theta;k) \Bigg] \Bigg\} \quad (6\text{-}25)$$

and we get

$$\frac{d^2}{d\theta^2}[\ln f_{X_I,X_Q}(\mathbf{X}_I,\mathbf{X}_Q)]\bigg|_{\theta=0} = \frac{\frac{d^2}{d\theta^2}[f_{X_I,X_Q}(\mathbf{X}_I,\mathbf{X}_Q)]\bigg|_{\theta=0}}{f_{X_I,X_Q}(\mathbf{X}_I,\mathbf{X}_Q)\bigg|_{\theta=0}}$$

$$= E_\psi \Bigg\{ \sum_{i=1}^{n} \frac{f''(0;i)}{f(0;i)} + \sum_{i=1}^{n} \sum_{\substack{j=1 \\ j \neq i}}^{n} \frac{f'(0;i)}{f(0;i)} \frac{f'(0;j)}{f(0;j)} \Bigg\} \quad (6\text{-}26)$$

Let us consider the first term in the above result. We have

$$f''(0;i) = s_i \mathbf{a}^T \mathbf{a} s_i^T f_{IQ}^{ii}(i) + s_i \mathbf{b}^T \mathbf{b} s_i^T f_{IQ}^{qq}(i)$$
$$+ s_i \mathbf{a}^T \mathbf{b} s_i^T f_{IQ}^{iq}(i) + s_i \mathbf{b}^T \mathbf{a} s_i^T f_{IQ}^{iq}(i) \quad (6\text{-}27)$$

Upon taking the expectation the cross-terms above vanish, leaving

$$E_\psi \left\{ \frac{f''(0;i)}{f(0;i)} \right\} = \frac{1}{2} v_i^2 \frac{\nabla^2 f_{IQ}(X_{Ii},X_{Qi})}{f_{IQ}(X_{Ii},X_{Qi})} \quad (6\text{-}28)$$

so that

$$E_\psi \left\{ \sum_{i=1}^{n} \frac{f''(0;i)}{f(0;i)} \right\} = \frac{1}{2} \sum_{i=1}^{n} v_i^2 \left[\frac{h''(R_i)}{h(R_i)} + \frac{h'(R_i)}{R_i h(R_i)} \right] \quad (6\text{-}29)$$

Here $v_i^2 = s_{Ii}^2 + s_{Qi}^2$, the square of the sampled *amplitude* term $v(t)$ of the signal.

Turning now to the second term in (6-26), we obtain

$$f'(0;i)f'(0;j) = [s_i \mathbf{a}^T f_{IQ}^i(i) + s_i \mathbf{b}^T f_{IQ}^q(i)]$$
$$\cdot [s_j \mathbf{a}^T f_{IQ}^i(j) + s_j \mathbf{b}^T f_{IQ}^q(j)] \quad (6\text{-}30)$$

so that, collecting terms in $\cos\psi$ and $\sin\psi$ separately, we get

$$f'(0;i)f'(0;j) = (\mathbf{s}_i \mathbf{l}_i^T \cos\psi + \mathbf{s}_i \mathbf{k}_i^T \sin\psi)(\mathbf{s}_j \mathbf{l}_j^T \cos\psi + \mathbf{s}_j \mathbf{k}_j^T \sin\psi) \tag{6-31}$$

where
$$\mathbf{l}_j = [f_{IQ}^i(j), f_{IQ}^i(j)] \tag{6-32}$$
and
$$\mathbf{k}_j = [-f_{IQ}^i(j), f_{IQ}^i(j)] \tag{6-33}$$

This gives

$$E_\psi \left\{ \sum_{\substack{i=1 \\ j \neq i}}^n \sum_{j=1}^n \frac{f'(0;i)}{f(0;i)} \frac{f'(0;j)}{f(0;j)} \right\}$$

$$= \frac{1}{2} \sum_{\substack{i=1 \\ j \neq i}}^n \sum_{j=1}^n \left[\mathbf{s}_i \mathbf{X}_i^T \mathbf{s}_j \mathbf{X}_j^T \tilde{g}_{LO}(R_i) \tilde{g}_{LO}(R_j) \right.$$
$$\left. + \mathbf{s}_i \mathbf{Y}_i^T \mathbf{s}_j \mathbf{Y}_j^T \tilde{g}_{LO}(R_i) \tilde{g}_{LO}(R_j) \right]$$

$$= \frac{1}{2} \left[\sum_{i=1}^n \tilde{g}_{LO}(R_i) \mathbf{s}_i \mathbf{X}_i^T \right]^2 + \frac{1}{2} \left[\sum_{i=1}^n \tilde{g}_{LO}(R_i) \mathbf{s}_i \mathbf{Y}_i^T \right]^2$$

$$- \frac{1}{2} \sum_{i=1}^n \tilde{g}_{LO}^2(R_i) \left[(\mathbf{s}_i \mathbf{X}_i^T)^2 + (\mathbf{s}_i \mathbf{Y}_i^T)^2 \right] \tag{6-34}$$

where
$$\mathbf{X}_i = (X_{Ii}, X_{Qi}) \tag{6-35}$$
and
$$\mathbf{Y}_i = (-X_{Qi}, X_{Ii}) \tag{6-36}$$

Now we have

$$(\mathbf{s}_i \mathbf{X}_i^T)^2 + (\mathbf{s}_i \mathbf{Y}_i^T)^2 = \mathbf{s}_i (\mathbf{X}_i^T \mathbf{X}_i + \mathbf{Y}_i^T \mathbf{Y}_i) \mathbf{s}_i^T$$

$$= \mathbf{s}_i \begin{bmatrix} R_i^2 & 0 \\ 0 & R_i^2 \end{bmatrix} \mathbf{s}_i^T$$

$$= R_i^2 v_i^2 \tag{6-37}$$

so that finally we are able to write our LO test statistic as

$$\lambda_{LO}(\mathbf{X}_I, \mathbf{X}_Q) = \frac{d^2}{d\theta^2} [\ln f_{\mathbf{X}_I, \mathbf{X}_Q}(\mathbf{X}_I, \mathbf{X}_Q)]\Big|_{\theta=0}$$

$$= \frac{1}{2} \sum_{i=1}^{n} v_i^2 \, \tilde{l}_{LO}(R_i) + \frac{1}{2} \left[\sum_{i=1}^{n} \tilde{g}_{LO}(R_i) \, s_i \, \mathbf{X}_i^T \right]^2$$

$$+ \frac{1}{2} \left[\sum_{i=1}^{n} \tilde{g}_{LO}(R_i) \, s_i \, \mathbf{Y}_i^T \right]^2 \quad (6\text{-}38)$$

where

$$\tilde{l}_{LO}(r) = \frac{h''(r)}{h(r)} - \left[\frac{h'(r)}{h(r)}\right]^2 + \frac{h'(r)}{rh(r)} \quad (6\text{-}39)$$

and $\tilde{g}_{LO}(r)$ is, as before,

$$\tilde{g}_{LO}(r) = \frac{-h'(r)}{rh(r)} \quad (6\text{-}40)$$

Note that $\tilde{g}_{LO}$ may be expressed as

$$\tilde{g}_{LO}(r) = -\frac{1}{r} \frac{d}{dr} \ln h(r) \quad (6\text{-}41)$$

and an alternate form for $\tilde{l}_{LO}(r)$ is

$$\tilde{l}_{LO}(r) = \frac{d^2}{dr^2} \ln h(r) + \frac{1}{r} \frac{d}{dr} \ln h(r) \quad (6\text{-}42)$$

Structure of the LO Detector

From the result of (6-38) we find that the LO detector for the detection of an incoherent narrowband signal in additive circularly symmetric narrowband noise is based on a test statistic consisting of two distinct components. The first term in $\lambda_{LO}(\mathbf{X}_I, \mathbf{X}_Q)$ of (6-38) is a *generalized envelope correlator* (GEC) statistic. It is formed as a correlation of the squared envelopes v_i^2 of the signal with nonlinearly transformed envelope (or squared-envelope) values $\tilde{l}_{LO}(R_i)$ of the input observation process. This term was not present in the LO test statistic of (5-27) for completely known, or coherent, narrowband signal detection. Notice that this term is present even when the signal has no amplitude modulation, so that the v_i^2 are all equal to some constant value. The second component of $\lambda_{LO}(\mathbf{X}_I, \mathbf{X}_Q)$ can be identified as the *square-law envelope detected* output of *quadrature generalized narrowband correlators*.

For Gaussian noise the noise envelope has the Rayleigh distribution, and we find that the first component in (6-38) is not a function of the R_i, the function $\tilde{l}_{LO}(r)$ being $-2/\sigma^2$, where σ^2 is the variance of the in-phase and quadrature noise components. In the Gaussian case the function $\tilde{g}_{LO}(r)$ is also just the constant $1/\sigma^2$, and the LO detector becomes the familiar bandpass matched filter and envelope detector combination.

Suppose, on the other hand, that the noise envelope has a density function $2\pi r h(r)$ in which h is described by h_k of (5-43). In this case we find that

$$\tilde{l}_{LO}(r) = \frac{-\sqrt{3}}{\sigma}\frac{1}{r} \tag{6-43}$$

and

$$\tilde{g}_{LO}(r) = \frac{\sqrt{3}}{\sigma}\frac{1}{r} \tag{6-44}$$

so that both the GEC component and the quadrature GNC component of the test statistic (6-38) are present. The latter component is a quadrature HNC test statistic. The GEC component gives little weighting to large envelope values, but small envelope values are given a large negative weighting. This means that for this noise model large envelope observations are not strongly indicative of the presence of a weak signal, whereas small envelope values are indicative of the absence of signal.

A convenient representation of the test statistic $\lambda_{LO}(\mathbf{X}_I, \mathbf{X}_Q)$ is obtained if we define *complex* signal and observation values

$$\tilde{s}_i = s_{Ii} + js_{Qi} \tag{6-45}$$
$$\tilde{X}_i = X_{Ii} + jX_{Qi} \tag{6-46}$$

Then we have $R_i^2 = |\tilde{X}_i|^2$ and $v_i^2 = |\tilde{s}_i|^2$ and the test statistic of (6-38) becomes

$$\lambda_{LO}(\tilde{\mathbf{X}}) = \frac{1}{2}\sum_{i=1}^{n}|\tilde{s}_i|^2 \tilde{l}_{LO}(|\tilde{X}_i|)$$
$$+ \frac{1}{2}\left|\sum_{i=1}^{n}\tilde{g}_{LO}(|\tilde{X}_i|)\tilde{s}_i \tilde{X}_i^*\right|^2 \tag{6-47}$$

Figure 6.1 is a block diagram of the LO detector implementing the test statistic $2\lambda_{LO}(\tilde{\mathbf{X}})$.

Before we proceed let us note that we could easily have included a random amplitude factor A in our signal component in (6-1), making the signal component $\theta A v(t) \cos[\omega_0 t + \phi(t) + \psi]$. In deriving the LO detector we would have had to perform a further averaging with respect to this random amplitude. The usual assumption is that A and ψ are independent random variables. By including the term A in the definition of **a** and **b** in (6-6) and (6-7), we would simply have obtained an additional factor $E\{A^2\}$ in the result for $\lambda_{LO}(\mathbf{X}_I, \mathbf{X}_Q)$ of (6-38), so that no essential difference is obtained. This case of a random but constant amplitude for the signal in the observation interval is called the *slow-fading* case. In the *fast-fading* case we let the amplitudes A_i for $i = 1,2,\ldots,n$ be i.i.d. random variables, independent of the phase ψ. In this case also the result for the LO statistic is easy to obtain; we leave its derivation as an exercise (Problem 6.1).

The locally optimum detection of incoherent narrowband signals in white noise has been considered by Antonov [1967b], Nirenberg [1975], Nirenberg and Middleton [1975], and Spaulding and Middleton [1977b]. These investigations did not specifically look into the case where the noise was also a narrowband process, but assumed it to be white and obtained detectors based directly on uniformly sampled values of the observation process. Lu and Eisenstein [1981] have considered the LO detector for the case of circularly symmetric in-phase and quadrature noise component distributions as we have, and they get similar results.

Asymptotically Optimum Detection and ARE

Consider the normalized equivalent statistic obtained by dividing $\lambda_{LO}(\tilde{\mathbf{X}})$ of (6-47) by n^2. For $n \to \infty$ we find that

$$\frac{1}{n^2} \lambda_{LO}(\tilde{\mathbf{X}}) \to \frac{1}{2} \left| \frac{1}{n} \sum_{i=1}^{n} \tilde{g}_{LO}(|\tilde{X}_i|) \tilde{s}_i \tilde{X}_i^* \right|^2 \quad (6\text{-}48)$$

indicating that for large n we may drop the GEC term from the LO detector test statistic. To find the efficacy of the LO detector we may therefore concentrate on the AO statistic

$$\lambda_{AO}(\tilde{\mathbf{X}}) = \frac{1}{2} \left| \sum_{i=1}^{n} \tilde{g}_{LO}(|\tilde{X}_i|) \tilde{s}_i \tilde{X}_i^* \right|^2 \quad (6\text{-}49)$$

Now this statistic can be expressed in the general form of (1-58), from which we can derive its *generalized efficacy*, as explained in Section 1.6 of Chapter 1. A direct approach to comparing the efficiencies of two quadrature generalized narrowband correlator/envelope detectors, each based on the use of some

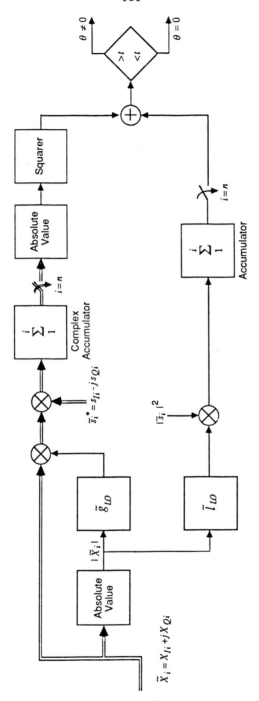

Figure 6.1 Locally Optimum Detector for Incoherent Signal in Narrowband Noise with Circularly Symmetric Probability Density Function

nonlinearity $\tilde{g}$ in place of $\tilde{g}_{LO}$ in (6-49) is to proceed from first principles as we did initially in Chapter 2. For this we need to establish that the distribution of $\lambda_{AO}(\tilde{X})$ is asymptotically a scaled χ^2-distribution with two degrees of freedom. This can be established conditionally for any value of the absolute phase ψ, and the distribution is independent of ψ. The ARE of two detectors based on different characteristics $\tilde{g}$ can then be found by expressing their asymptotic power function in terms of their noncentrality parameters under the null and asymptotic alternative hypotheses. We will not develop the details here, but leave it as an exercise for the reader to show that the ARE of two detectors based on two different characteristics $\tilde{g}_1$ and $\tilde{g}_2$ for this incoherent signal detection problem is the same as the ARE for the corresponding GNC detectors in the case of completely known narrowband signals (Problem 6.2).

Asymptotically optimum detection of narrowband signals has also been addressed by Levin and Kushnir [1969] and by Levin and Rybin [1969]. While we do not do so here, it is possible to obtain the explicit form of the locally optimum Bayes detector for this detection problem of an incoherent signal, as we did for the coherent case in Chapter 5. Locally optimum Bayes detection has been considered for incoherent signals by Spaulding and Middleton [1977b] and Maras, Davidson, and Holt [1985a], although in both places directly sampled bandpass observation waveforms are used rather than samples of the in-phase and quadrature components.

Izzo and Paura [1986] have considered asymptotically optimum detection of incoherent narrowband signals for diversity reception, in which the observations at a number of different locations with independent noise processes and random amplitudes are utilized. Their result reduces to the one we have developed above for the special case of one-channel reception. Other recent investigations of locally optimum detection of incoherent signals in non-Gaussian noise include those of Maras, Davidson, and Holt [1985b] and of Lu and Eisenstein [1981, 1983].

6.3 Detection of a Noncoherent Pulse Train

We shall now consider as a special case of narrowband signal detection the detection of a *pulse train* in additive noise. This type of signal is clearly of interest in radar and sonar detection systems, as well as in communication systems employing RF pulses for signaling.

6.3.1 The Observation Model

Any *single* RF signal pulse in a narrowband pulse train will be modeled as giving rise to an observation waveform of the form of (6-1). The amplitude and phase modulations $v(t)$ and $\phi(t)$ now describe the basic pulse characteristic. In addition, we can now allow the random phase ψ to be different for each pulse, and also allow a non-uniform amplitude characteristic for the train of pulses. This leads to the following model for the received waveform:

$$X(t) = \theta \, \text{Re}\{\tilde{V}(t) e^{j[\omega_0 t + \psi(t)]}\} + W(t), \quad 0 \leq t \leq n\Delta \tag{6-50}$$

where $\tilde{V}(t)$ is the known complex low-pass pulse train envelope given by

$$\tilde{V}(t) = \sum_{i=1}^{n} e_i v_i(t) e^{j \phi_i(t)}$$

$$= \sum_{i=1}^{n} e_i v_i(t) \, e^{j \sum_{i=1}^{n} \phi_i(t)}, \quad 0 \leq t \leq n\Delta \tag{6-51}$$

with

$$v_i(t) = p[t - (i-1)\Delta], \quad i = 1,2,\ldots,n \tag{6-52}$$

and

$$\phi_i(t) = \phi[t - (i-1)\Delta], \quad i = 1,2,\ldots,n \tag{6-53}$$

The basic pulse amplitude and phase characteristics $p(t)$ and $\phi(t)$ are assumed to be non-zero only within $(0,\Delta)$. The e_i in (6-51) are known relative amplitude values of the pulses in the train. It is useful to allow this generalization, since in many applications a received train can have some known amplitude weighting; for example, we may be detecting a PAM signal or radar returns which have been amplitude modulated by the receiving scanning antenna pattern.

If we assume that the random phase characteristic $\psi(t)$ is not a function of t (that is, it is a random variable) we have the *coherent pulse train* detection problem. Notice that in our previous terminology *each* pulse is an *incoherent* narrowband pulse if we assume that ψ is a uniformly distributed random variable, which will be our standard assumption. More generally, we may assume that $\psi(t)$ is a constant for each pulse but may be different from pulse to pulse. In this case we can define

$$\psi(t) = \psi_i, \quad (i-1)\Delta < t \leq i\Delta \tag{6-54}$$

If the random, uniformly distributed phases ψ_i are a set of *independent* phases, we get the *noncoherent pulse train* detection problem.

We notice that the detection of a *coherent* pulse train with uniformly distributed absolute phase is simply a special case of the problem we have already considered in Section 6.2. Thus we will be primarily interested here in the *noncoherent* pulse train problem. Suppose that we are dealing with a situation in which the width, d, of each pulse is narrow relative to the period Δ, and in which the bandwidth of the noise is sufficiently large to allow us to treat each pulse period independently. We have discussed in some detail a similar assumption for the low-pass pulse train case in Section 2.2 of Chapter 2. Then in principle we can simply apply the results we have already obtained above for what was effectively the single-pulse detection problem (based on sampled observations within one pulse period); the locally optimum scheme may be obtained by adding an integrator or accumulator to the output producing the test statistic of the single-pulse detector of Figure 6.1.

As an alternative, we may *assume* that our detector is now presented with the $2n$ outputs of a quadrature matched-filter matched to the basic pulse $v(t)\cos[\omega_0 t + \phi(t)]$. For Gaussian noise this will, under the independence assumption above [see also Section 2.2 of Chapter 2], give the optimum scheme. Indeed, we have observed that in the single-pulse case for Gaussian noise the test statistic of (6-38) is simply the square of the output envelope of a bandpass matched filter, or the output of a quadrature matched-filter followed by a square-law-envelope detector. Such a linear matched filter is simple to implement and is a very frequently encountered processing element at the receiver input. If the input noise is not Gaussian the output will also be non-Gaussian, and it should then be possible to perform an appropriate nonlinear integration of the matched filter outputs. This is what we will now study. We should keep in mind, however, that the constraint of a linear pre-filter can lead to a significant compromise of the detector's ability to deal with impulsive noise.

Let us denote by Z_{Ii} and Z_{Qi} the in-phase and quadrature components of the output of a quadrature matched filter matched to the basic pulse $p(t)\cos[\omega_0 t + \phi(t)]$. The index i now indicates the outputs at the end of the i-th pulse period. The in-phase components Z_{Ii} may be described by

$$Z_{Ii} = \int_0^\Delta p(t) \cos[\omega_0 t + \omega_0(i-1)\Delta + \phi(t)] X[t + (i-1)\Delta] \, dt$$

$$= \theta e_i \int_0^\Delta p^2(t) \cos[\omega_0 t + \omega_0(i-1)\Delta + \phi(t)]$$

$$\cdot \cos[\omega_0 t + \omega_0(i-1)\Delta + \phi(t) + \psi_i]\, dt$$

$$+ \int_0^\Delta p(t) \cos[\omega_0 t + \omega_0(i-1)\Delta + \phi(t)]\, W[t + (i-1)\Delta]\, dt \quad (6\text{-}55)$$

Assuming that $\omega_0 \gg 2\pi/d$, (where d is the pulse width), using the normalization

$$\int_0^\Delta p^2(t)\, dt = 2 \quad (6\text{-}56)$$

and the definition

$$U_{Ii} = \int_0^\Delta p(t) \cos[\omega_0 t + \omega_0(i-1)\Delta + \phi(t)]\, W[t + (i-1)\Delta]\, dt \quad (6\text{-}57)$$

we get

$$Z_{Ii} = \theta e_i \cos \psi_i + U_{Ii} \quad (6\text{-}58)$$

Similarly, we have

$$Z_{Qi} = \int_0^\Delta p(t) \sin[\omega_0 t + \omega_0(i-1)\Delta + \phi(t)]\, X[t + (i-1)\Delta]\, dt$$

$$= -\theta e_i \sin \psi_i + U_{Qi} \quad (6\text{-}59)$$

where

$$U_{Qi} = \int_0^\Delta p(t) \sin[\omega_0 t + \omega_0(i-1)\Delta + \phi(t)]\, W[t + (i-1)\Delta]\, dt \quad (6\text{-}60)$$

If $W(t)$ is zero-mean bandpass noise with a constant power spectral density $N_0/2$ over a band of frequencies of width large compared to the signal bandwidth, we find that U_{Ii} and U_{Qi} are zero-mean and uncorrelated random variables with variances $N_0/2$. In addition, this condition will allow us to assume that the sequence $\{(U_{Ii}, U_{Qi}), i = 1,2,...,n\}$ of bivariate noise components is an i.i.d. sequence. We will assume, as we have been doing, that (U_{Ii}, U_{Qi}) has a *circularly symmetric* probability density function $f_{IQ}(u,v) = h(\sqrt{u^2 + v^2})$. It is also possible to obtain some more general results without this assumption, as we will see.

Note that we would end up with the same model as described by (6-58) and (6-59) even if the phase term $\omega_0(i-1)\Delta$ were to be dropped from the argument of the matched filter impulse responses; this is because ψ_i is uniformly distributed on $[0,2\pi]$. Alternatively, we could have assumed that $\omega_0(i-1)\Delta$ is a multiple of 2π, without much loss of generality. The phases $\omega_0(i-1)\Delta$ could also have been dropped if our original model of (6-50) had been modified so that the pulse train is made up of a single RF pulse repeated with period Δ and random phases ψ_i. In all these cases our basic model for the Z_{Ii} and Z_{Qi} remains the same.

We find that the model describing the components Z_{Ii} and Z_{Qi} of the quadrature single-pulse matched filter outputs is mathematically very similar to that for the sampled in-phase and quadrature components X_{Ii} and X_{Qi} of (6-4) and (6-5). The main difference is that for the noncoherent pulse train we find that the ψ_i are i.i.d. uniformly distributed random variables. Otherwise, we get Z_{Ii} and Z_{Qi} from X_{Ii} and X_{Qi} of (6-4) and (6-5), respectively, by letting $s_{Ii} = e_i$ and $s_{Qi} = 0$.

6.3.2 The Locally Optimum Detector

To find the LO detector based on the $\mathbf{Z}_I = (Z_{I1}, Z_{I2}, \ldots, Z_{In})$ and $\mathbf{Z}_Q = (Z_{Q1}, Z_{Q2}, \ldots, Z_{Qn})$ we can use our earlier result by making a simple observation. This is that now *individually* the bivariate samples (Z_{Ii}, Z_{Qi}) are modeled by (6-4) and (6-5), with $s_{Ii} = e_i$ and $s_{Qi} = 0$. Thus given only *one* bivariate observation (Z_{Ii}, Z_{Qi}), the LO test statistic would be

$$\lambda_{LO}(Z_{Ii}, Z_{Qi}) = \frac{1}{2} e_i^2 \left[\frac{h''(R_i)}{h(R_i)} + \frac{h'(R_i)}{R_i h(R_i)} \right] \qquad (6\text{-}61)$$

where $R_i = \sqrt{Z_{Ii}^2 + Z_{Qi}^2}$. This follows from (6-47) which for one observation becomes

$$\lambda_{LO}(\tilde{X}_i) = \frac{1}{2} e_i^2 \, \tilde{l}_{LO}(R_i) + \frac{1}{2} e_i^2 R_i^2 \tilde{g}_{LO}^2(R_i) \qquad (6\text{-}62)$$

together with (6-39) and (6-40). Now since the pairs (Z_{Ii}, Z_{Qi}) are *independently distributed* for $i = 1, 2, \ldots, n$ under both hypotheses we are immediately able to write down the result

$$\begin{aligned}\lambda_{LO}(\mathbf{Z}_I, \mathbf{Z}_Q) &= \frac{1}{2} \sum_{i=1}^{n} e_i^2 \left[\frac{h''(R_i)}{h(R_i)} + \frac{h'(R_i)}{R_i h(R_i)} \right] \\ &= \frac{1}{2} \sum_{i=1}^{n} e_i^2 [\tilde{l}_{LO}(R_i) + R_i^2 \tilde{g}_{LO}^2(R_i)] \qquad (6\text{-}63)\end{aligned}$$

It is seen that the locally optimum detector for this pulse-train detection problem is a *generalized envelope correlator* detector. Notice that one need not obtain Z_{Ii} and Z_{Qi} separately since only the envelope value is used. Thus it is possible to use a bandpass matched filter followed by an envelope detector to obtain the R_i, instead of the quadrature matched filter structure for the Z_{Ii} and Z_{Qi}. In the case of Gaussian noise, for which

$$h(r) = \frac{1}{2\pi\sigma^2} e^{-r^2/2\sigma^2}, \quad r \geq 0 \tag{6-64}$$

where $\sigma^2 = N_0/2$, we obtain the LO statistic

$$\lambda_{LO}(\mathbf{R}) = \frac{1}{\sigma^2} \sum_{i=1}^{n} e_i^2 \left[\frac{R_i^2}{2\sigma^2} - 1 \right] \tag{6-65}$$

which is equivalent to the *square-law-envelope correlator* statistic

$$T_{SEC}(\mathbf{R}) = \sum_{i=1}^{n} e_i^2 R_i^2 \tag{6-66}$$

Here $\mathbf{R} = (R_1, R_2, \ldots, R_n)$. Suppose, on the other hand, that we have

$$h(r) = \frac{1}{2\pi c^2} e^{-r/c}, \quad r \geq 0 \tag{6-67}$$

for some positive constant c. Then we find

$$\frac{h''(r)}{h(r)} + \frac{h'(r)}{rh(r)} = \frac{1}{c^2} - \frac{1}{cr} \tag{6-68}$$

and the equivalent LO detector test statistic is the generalized envelope correlator statistic

$$T(\mathbf{R}) = \sum_{i=1}^{n} -\frac{e_i^2}{R_i} \tag{6-69}$$

These two test statistics are quite different, which indicates that the detector based on $T_{SEC}(\mathbf{R})$ may become quite poor in its performance relative to an LO detector for non-Gaussian noise. Notice that $T(\mathbf{R})$ of (6-69) does have the characteristic that its value tends to be larger when the e_i^2 and R_i are correlated. From Figure 6.2(a) we see that a type of envelope limiter nonlinear characteristic is suggested for a reasonable sub-optimum GEC test statistic for this case. Figure 6.2(b) shows that such a characteristic should provide acceptable performance in the Gaussian case also, so that it would probably be a good compromise characteristic for both Rayleigh and heavier-tailed noise envelope densities.

The result of (6-63) for the LO detector test statistic is applicable specifically when the in-phase and quadrature components U_{Ii} and U_{Qi} of the noise have a circularly symmetric bivariate density function, with $2\pi r h(r)$ the density of their envelope. Upon tracing the development of the LO test statistic in Section 6.2 for the special case of a single observation pair (X_{Ii}, X_{Qi}), it becomes quite clear that without circular symmetry being assumed the LO detector test statistic in the present case becomes

$$\lambda_{LO}(Z_I, Z_Q) = \frac{1}{2} \sum_{i=1}^{n} e_i^2 \frac{\nabla^2 f_{IQ}(Z_{Ii}, Z_{Qi})}{f_{IQ}(Z_{Ii}, Z_{Qi})} \tag{6-70}$$

It is also possible to consider a pulse-to-pulse independent *amplitude fluctuation* by including random amplitudes A_i in our model of (6-58) and (6-59). Specifically, we may replace the e_i by $A_i e_i$, where the A_i, $i = 1,2,...,n$ form a sequence of i.i.d. random variables. We can then conclude, again based on our discussion of the amplitude fading situation in Section 6.2, that the structure of the LO detector does not change. Of course, in the non-local case the detector test statistic is influenced by the fading amplitude distribution. For example, it is well-known that the square-law envelope detector is *optimum* for Gaussian noise and Rayleigh fading, whereas for no amplitude fading in the Gaussian case the optimum detector uses a linear weighting for large envelope observations [Helstrom, 1968].

Efficacy and ARE Comparisons

The above results for the LO test statistic for our problem suggest that we consider as a useful class of test statistics the generalized correlator statistic operating on the *complex* data $\tilde{Z}_i = Z_{Ii} + j Z_{Qi}$, $i = 1,2,...,n$:

$$T_{GC}(\tilde{Z}) = \sum_{i=1}^{n} e_i^2 g(Z_{Ii}, Z_{Qi}) \tag{6-71}$$

Let us now derive the efficacy of this test statistic for our detection problem. We obtain

$$E\{g(Z_{Ii},Z_{Qi})\}$$
$$= E\{\int_{-\infty}^{\infty}\int_{-\infty}^{\infty} g(u,v) f_{IQ}(u - \theta e_i \cos\psi_i, v + \theta e_i \sin\psi_i) du dv\} \quad (6\text{-}72)$$

from which it can be easily deduced that the first derivative of the expected value with respect to θ is zero at $\theta = 0$. Thus we consider the second derivative, getting

$$\frac{d^2}{d\theta^2} E\{g(Z_{Ii},Z_{Qi})\}\bigg|_{\theta=0} = \frac{1}{2} e_i^2 \int_{-\infty}^{\infty}\int_{-\infty}^{\infty} g(u,v) \nabla^2 f_{IQ}(u,v) du\, dv \quad (6\text{-}73)$$

Assuming without loss of generality that $E\{g(Z_{Ii},Z_{Qi})\}$ is zero for $\theta = 0$, we get for the variance when $\theta = 0$

$$V\{g(Z_{Ii},Z_{Qi})\} = \int_{-\infty}^{\infty}\int_{-\infty}^{\infty} g^2(u,v) f_{IQ}(u,v) du\, dv \quad (6\text{-}74)$$

Using our general result of Section 1.6, we obtain the efficacy ξ_{GC} of the test statistic of (6-71) for our problem to be

$$\xi_{GC} = \frac{1}{16} \frac{\left[\int_{-\infty}^{\infty}\int_{-\infty}^{\infty} g(u,v)\nabla^2 f_{IQ}(u,v) du\, dv\right]^2}{\int_{-\infty}^{\infty}\int_{-\infty}^{\infty} g^2(u,v) f_{IQ}(u,v) du\, dv} \lim_{n\to\infty} \frac{1}{n}\sum_{i=1}^{n} e_i^4 \quad (6\text{-}75)$$

Assuming that the limit above is finite and positive, we get for the normalized efficacy the result

$$\tilde{\xi}_{GC} = \frac{1}{16} \frac{\left[\int_{-\infty}^{\infty}\int_{-\infty}^{\infty} g(u,v)\nabla^2 f_{IQ}(u,v) du\, dv\right]^2}{\int_{-\infty}^{\infty}\int_{-\infty}^{\infty} g^2(u,v) f_{IQ}(u,v) du\, dv} \quad (6\text{-}76)$$

Then, applying the Schwarz inequality, we find that $\tilde{\xi}_{GC}$ is maximized when g is the LO characteristic $\nabla^2 f_{IQ}/f_{IQ}$, for which the efficacy becomes

$$\tilde{\xi}_{LO} = \int_{-\infty}^{\infty}\int_{-\infty}^{\infty} \left[\frac{1}{4}\frac{\nabla^2 f_{IQ}(u,v)}{f_{IQ}(u,v)}\right]^2 f_{IQ}(u,v) du\, dv \quad (6\text{-}77)$$

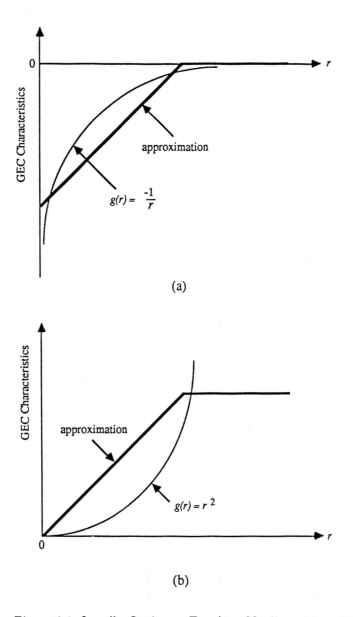

Figure 6.2 Locally Optimum Envelope Nonlinearities and Approximations for (a) Circularly Symmetric Exponential Noise Envelope Density, (b) Rayleigh Noise Envelope Density (Gaussian Noise)

It is instructive to compare this result with the definition of Fisher's information for location $I(f)$ of (2-12) for known-signal detection.

In the special case of circular symmetry with $f_{IQ}(u,v) = h(r)$ and $r = \sqrt{u^2 + v^2}$, and when $g(u,v)$ can be similarly written as $\tilde{g}(r)$, the normalized efficacy becomes

$$\tilde{\xi}_{GC} = \frac{\pi}{8} \frac{\left[\int_0^\infty \tilde{g}(r) \left[\frac{h''(r)}{h(r)} + \frac{h'(r)}{rh(r)} \right] h(r) r \, dr \right]^2}{\int_0^\infty \tilde{g}^2(r) h(r) r \, dr} \qquad (6\text{-}78)$$

These results allow a comparison to be made of detectors employing different characteristics g or $\tilde{g}$ in various noise environments.

Some details of such comparisons have been given in a recent paper [Cimini and Kassam, 1983]. We will not discuss further the asymptotic performance characteristics of the generalized correlator detectors (operating on in-phase and quadrature data treated as complex values) for our noncoherent pulse train detection problem. To complete our treatment we will now continue on to examine quantization of the data in this detection problem.

Before continuing let us comment on the factor $1/16$ in the result of (6-75) for ξ_{GC}. This arises instead of $1/4$, which would be obtained if our usual definition of efficacy had been applied because here we have taken the *second* derivative at the origin in (6-73). Another way to view this result is that our signal strength parameter is now θ^2, and the first derivative of the mean function with respect to θ^2, at $\theta^2 = 0$, is one-half the value of its second derivative with respect to θ. This will also be clarified in the next chapter. If, on the other hand, we note that $E\{\theta^2 \cos^2 \psi_i\} = E\{\theta^2 \sin^2 \psi_i\} = \theta^2/2$ and use $\theta^2/2$ as the signal strength parameter, the factor $1/16$ in (6-75) is reduced to $1/4$. Thus one has to be consistent in applying the efficacy to obtain approximate large-sample-size detection performance; the same definition of signal strength must be used throughout. Similarly, in ARE comparisons one gets correct results as long as consistent use in made of one definition for the signal strength parameter. Finally note that the quantity $(1/4) \nabla^2 f_{IQ}(u,v)/f_{IQ}(u,v)$ in (6-77) can be viewed as that scaled version of the LO characteristic which arises as the small-θ^2 approximation of the likelihood ratio. When interpreting $\theta^2/2$ as the signal parameter we may, for consistency, scale up the above LO characteristic by a factor of 2.

6.4 Asymptotically Optimum Quantization

In Section 6.2 above we have obtained the asymptotically optimum test statistic for detection of an incoherent narrowband signal in additive noise. Although we did not derive it explicitly, we indicated how the efficacy of any such quadrature GNC-envelope detector statistic can be obtained. The result is the same as that obtained for the completely known narrowband signal detection problem in Chapter 5. This means that we can consider quantization of the envelope nonlinearity in the test statistic of (6-49) exactly as we did earlier for the coherent or completely known narrowband signal detection problem.

It is of even greater practical significance to consider optimum quantization in the test statistic of the LO detector for our noncoherent pulse train detection problem. In radar receivers the envelope detected output of a bandpass matched filter are often integrated digitally, meaning that the correlation in (6-63) is performed on quantized envelope values $\tilde{q}(R_i)$. Thus for both Gaussian noise and non-Gaussian noise it is useful to know the effect of quantization and to consider optimization of such quantization. We shall show below that in many cases one can get very good performance by employing only a few quantization levels.

6.4.1 M-Interval Envelope Quantization

Let us first focus on the special but generally applicable case of circular symmetry for the bivariate density function of the in-phase and quadrature noise components U_{Ii} and U_{Qi} in our observation model of (6-58) and (6-59). In this case the LO test statistic is given by (6-63), which we may write as

$$\lambda_{LO}(\mathbf{R}) = \sum_{i=1}^{n} e_i^2 \tilde{p}_{LO}(R_i) \tag{6-79}$$

where

$$\tilde{p}_{LO}(R_i) = \frac{1}{2}[\tilde{l}_{LO}(R_i) + R_i^2 \tilde{g}_{LO}^2(R_i)] \tag{6-80}$$

We now want to replace $\tilde{p}_{LO}$ by an M-interval quantizer $\tilde{q}$, defined by

$$\tilde{q}(r) = l_j \text{ for } t_{j-1} < r \leq t_j, \ j = 1,2,...,M \tag{6-81}$$

with $t_0 \triangleq 0$ and $t_M \triangleq \infty$, $\mathbf{t} = (t_0, t_1, ..., t_M)$ being the vector of input breakpoints and the l_j being the output levels of the quantizer. The resulting test statistic

$$T_{QEC}(\mathbf{R}) = \sum_{i=1}^{n} e_i^2 \, \tilde{q}(R_i) \qquad (6\text{-}82)$$

is a *quantizer envelope correlator* (QEC) statistic. The detector based on this statistic is shown in Figure 6.3. We can find its efficacy by employing the result (6-78), which gives for the normalized efficacy the result

$$\tilde{\xi}_{QEC} = \frac{\pi}{2} \frac{\left[\sum_{j=1}^{M} l_j \int_{t_{j-1}}^{t_j} \tilde{p}_{LO}(r) h(r) r \, dr\right]^2}{\sum_{j=1}^{M} l_j^2 \int_{t_{j-1}}^{t_j} h(r) r \, dr}$$

$$= \frac{1}{4} \frac{\left[\sum_{j=1}^{M} l_j \int_{t_{j-1}}^{t_j} \tilde{p}_{LO}(r) f_E(r) \, dr\right]^2}{\sum_{j=1}^{M} l_j^2 \left[F_E(t_j) - F_E(t_{j-1})\right]} \qquad (6\text{-}83)$$

where f_E and F_E are the noise envelope density and distribution functions, respectively. It should be noted that this is a valid result provided that the quantized data $\tilde{q}(R_i)$ has a mean value zero under the null hypothesis.

We find that the result above is very similar to the one we obtained earlier for the efficacy of the QNC detector [Equation (5-67)] for detection of a coherent narrowband signal. Proceeding as we have done before, we find that the LO levels for a given breakpoint vector $\mathbf{t}$ are those maximizing $\tilde{\xi}_{QEC}$; one such set of levels are

$$l_j = \frac{\int_{t_{j-1}}^{t_j} \tilde{p}_{LO}(r) f_E(r) dr}{F_E(t_j) - F_E(t_{j-1})}, \quad j = 1, 2, \ldots, M \qquad (6\text{-}84)$$

Note that with these levels,

$$E\{\tilde{q}(R_i) \mid \theta = 0\} = \pi \sum_{j=1}^{M} \int_{t_{j-1}}^{t_j} [r h''(r) + h'(r)] \, dr$$

$$= \pi \sum_{j=1}^{M} t_j h'(t_j) - t_{j-1} h'(t_{j-1})$$

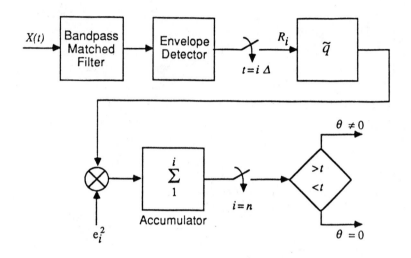

Figure 6.3 Quantizer Envelope Correlator Detector for Noncoherent Pulse Train in Additive Narrowband Noise with Circularly Symmetric Bivariate Noise Probability Density Function

$$= \pi \lim_{r \to \infty} [rh'(r)] - \pi \lim_{r \to 0} [rh'(r)] \qquad (6\text{-}85)$$

Provided that this value is zero, as it is for almost all cases of interest, we find that the quantization levels of (6-84) result in the quantized envelope values having zero means under the noise-only condition.

Use of the optimum levels of (6-84) for the quantization gives for the normalized efficacy the result

$$\tilde{\xi}_{QEC,LO} = \frac{1}{4} \sum_{j=1}^{M} \frac{\left[\int_{t_{j-1}}^{t_j} \tilde{p}_{LO}(r) f_E(r)\, dr \right]^2}{F_E(t_j) - F_E(t_{j-1})} \qquad (6\text{-}86)$$

We may then further maximize this efficacy with respect to the breakpoints vector t. Taking the derivative of $\tilde{\xi}_{QEC,LO}$ with respect to t_j and setting the result equal to zero gives the necessary conditions

$$\tilde{p}_{LO}(t_j) = \frac{1}{2} \left\{ \frac{\int_{t_{j-1}}^{t_j} \tilde{p}_{LO}(r) f_E(r)\, dr}{F_E(t_j) - F_E(t_{j-1})} + \frac{\int_{t_j}^{t_{j+1}} \tilde{p}_{LO}(r) f_E(r)\, dr}{F_E(t_{j+1}) - F_E(t_j)} \right\},$$
$$j = 1,2,...,M-1 \qquad (6\text{-}87)$$

This can also be written as

$$\tilde{p}_{LO}(t_j) = \frac{l_j + l_{j+1}}{2}, \quad j = 1,2,...,M-1 \qquad (6\text{-}88)$$

where the l_j are the LO levels given by (6-84). The $2M-1$ equations (6-84) and (6-88) can thus be solved simultaneously to get the AO quantizer parameters.

6.4.2 Optimum Envelope Quantization for Gaussian Noise

As a specific example, consider the case where the envelope of the noise has a Rayleigh distribution, so that

$$f_E(r) = \frac{r}{\sigma^2} e^{-r^2/2\sigma^2} \qquad (6\text{-}89)$$

$$F_E(r) = 1 - e^{-r^2/2\sigma^2} \qquad (6\text{-}90)$$

and

$$\tilde{p}_{LO}(r) = \frac{r^2}{2\sigma^4} - \frac{1}{\sigma^2} \qquad (6\text{-}91)$$

This gives for the LO levels

$$l_j = \frac{\int_{t_{j-1}}^{t_j} \frac{r^3}{2\sigma^6} e^{-r^2/2\sigma^2}\, dr}{e^{-t_{j-1}^2/2\sigma^2} - e^{-t_j^2/2\sigma^2}} - \frac{1}{\sigma^2}$$

$$= \frac{u_{j-1} e^{-u_{j-1}} - u_j e^{-u_j}}{\sigma^2 (e^{-u_{j-1}} - e^{-u_j})}, \quad j = 1,2,...,M \qquad (6\text{-}92)$$

where for convenience we have defined the u_j to be $t_j^2/2\sigma^2$. From

(6-88) we also get

$$\frac{t_j^2}{2\sigma^4} - \frac{1}{\sigma^2} = \frac{l_j + l_{j+1}}{2}, \quad j = 1,2,...,M-1 \qquad (6\text{-}93)$$

or in terms of $u_j = t_j^2/2\sigma^2$,

$$u_j = \sigma^2 \frac{l_j + l_{j+1}}{2} + 1, \quad j = 1,2,...,M-1 \qquad (6\text{-}94)$$

Equations (6-92) and (6-94) may be solved simultaneously for the l_j, σ^2 and the u_j quite easily. We can then find the ARE of the AO QEC detector relative to the SEC detector [Eq. (6-66)] which is locally optimum for this Gaussian noise problem. Table 6.1 shows some numerical results for the AO levels l_j and breakpoints t_j. The optimum levels can be scaled up or down by any constant, and so are independent of σ^2. Table 6.1 also shows the ARE obtained for various values of M using optimum quantizer parameters. The corresponding figures for the asymptotic loss are the *square roots* of the ARE, in dB. In Section 2.3 of Chapter 2 we discussed how the ARE is the ratio of the squares of signal strengths required by two detectors for their asymptotic performance to remain identical. In the present situation our measure of signal strengths is effectively θ^2; note that we obtained the efficacy by taking the *second* derivative of the expected value of the test statistic, and an equivalent result would have been obtained by taking the *first* derivative of the expected value with respect to θ^2, evaluated for the limiting case $\theta \to 0$. Thus the asymptotic loss shows how much additional SNR is required by the QEC detector to achieve the same level of asymptotic performance as that of the SEC detector. Table 6.1 shows that with four quantization intervals the asymptotic loss need only be about 0.25 dB.

While good asymptotic performance is indicated in the above example for a relatively small number of levels in the envelope quantization, the finite sample-size performance is of primary interest in a practical scheme. Now for any noise we have found that the LO detector is the same for both the pulse-to-pulse independent amplitude fluctuation case and the non-fluctuating amplitude case. For Gaussian noise this LO detector is the square-law envelope correlator detector with the test statistic of (6-66). For finite-sample-size analysis we can investigate both the amplitude-fluctuation and non-fluctuating-amplitude cases, and in the former case an assumption has to be made about the amplitude univariate density function.

M	$\{t_j/\sigma\}_{j=1}^{M-1}$	$\{l_j\}_{j=1}^{M}$	ARE	ASYMPTOTIC LOSS (dB)
2	1.785	-0.40 1.59	0.648	0.94
3	1.426 2.283	-0.58 0.61 2.61	0.820	0.43
4	1.225 1.880 2.595	-0.67 0.17 1.36 3.37	0.891	0.25
5	1.094 1.641 2.173 2.810	-0.73 -0.07 0.77 1.95 3.95	0.927	0.16
6	0.997 1.482 1.924 2.393 2.978	-0.77 -0.23 0.43 1.27 2.45 4.43	0.947	0.12

Table 6.1 Asymptotically Optimum Envelope Quantizer Parameters and ARE of the M-Interval QEC Detectors Relative to the SEC Detector for Gaussian Noise

A common parametric assumption has been that the amplitude of the signal has a Rayleigh distribution independent of its phase. For the amplitude-fluctuation situation our observations Z_{Ii} and Z_{Qi} will thus be described by

$$Z_{Ii} = \theta A_i \cos \psi_i + U_{Ii} \tag{6-95}$$

$$Z_{Qi} = -\theta A_i \sin \psi_i + U_{Qi} \tag{6-96}$$

where the A_i are i.i.d. Rayleigh random variables. We are assuming that $e_i = 1$ for our finite-sample-size results, so that all the pulses have a constant average amplitude level, to simplify the exact performance computation. Since the parameter θ controls the signal strength we may assume that the U_{I_i}, U_{Q_i}, are unit-variance and zero-mean noise components, in addition to the assumption that they are Gaussian (and independent) random variables. We also know that $A_i \cos \psi_i$ and $A_i \sin \psi_i$ are independent and zero-mean Gaussian random variables. The parameter $S = \theta^2 E\{A_i^2 \cos^2 \psi_i\} = \theta^2 E\{A_i^2\}/2$ is therefore the SNR. Now under the alternative hypothesis when $\theta \neq 0$ the Z_{I_i} and Z_{Q_i} are independent zero-mean Gaussian observations with variances $1 + S$, so that the envelope R_i in this case has the Rayleigh density

$$f_E(r) = \frac{r}{1+S} e^{-r^2/2(1+S)} \tag{6-97}$$

When $S = 0$ (noise-only) the above gives the distribution of the noise envelope. For this problem of a *scale change* in the Rayleigh density function of the noise envelope due to signal presence, the UMP detector is quite easily seen to be the *square-law detector* [the SEC detector with $e_i = 1$ in (6-66)].

In the non-fluctuating signal model we let the A_i all be equal to some *constant A*. The SNR is then $S = \theta^2 A^2/2$. For $\theta^2 > 0$, that is when the signal is present, the envelopes R_i of the observations have the *Rician* density function

$$f_E(r) = r \, e^{-(S + r^2/2)} I_0(r \sqrt{2S}) \tag{6-98}$$

where I_0 is the zeroth-order modified Bessel function. The *optimum* detector for this case uses the test statistic

$$T(\mathbf{R}) = \sum_{i=1}^{n} \ln I_0(R_i \sqrt{2S}) \tag{6-99}$$

as is quite easy to obtain from the likelihood ratio for the envelope vector $\mathbf{R} = (R_1, R_2, ..., R_n)$. Notice that the above statistic depends on S, and does not yield a UMP test. For small S, however, the approximation $\ln I_0(R_i \sqrt{2S}) \approx 2SR_i^2$ yields, as expected, the square-law envelope detector as the LO detector. On the other hand for large values of S the test statistic is usually approximated by the equivalent *linear* envelope detector statistic

$$T_{LE}(\mathbf{R}) = \sum_{i=1}^{n} R_i \tag{6-100}$$

In practice the linear envelope detector is often used when the low SNR assumption is not valid.

We have entered into the above discussion to explain why we will now use the square-law envelope detector based on the test statistic

$$T_{SE}(\mathbf{R}) = \sum_{i=1}^{n} R_i^2 \qquad (6\text{-}101)$$

in obtaining finite-sample-size performance comparisons for Rayleigh amplitude-fluctuating signal detection (in Gaussian noise), whereas we will use the linear envelope detector to obtain performance comparisons with our asymptotically optimum QEC scheme for the non-fluctuating signal model.

In obtaining numerical results for the detection probabilities it is very convenient to use integer-valued quantizer output levels. Since the performance of a QEC detector remains invariant to a linear transformation performed on all the levels l_j, we can generally find a set of integer-valued levels giving close-to-optimum asymptotic performance. From Table 6.1, we find that for $M = 3$ we can use levels $\{0, 1, 3\}$ for $\{l_1, l_2 l_3\}$, and for $m = 4$ a suitable set of levels is $\{0, 1, 2, 4\}$. For $m = 2$ we can, of course, use the levels 0 and 1. Using these levels and the asymptotically optimum breakpoints, the exact detection probabilities for a finite number n of observations can be obtained for both non-fluctuating and Rayleigh fluctuating amplitudes.

Table 6.2 summarizes the results of exact computation of the detection probabilities for the quantized envelope detector and its comparison with the unquantized linear and square-law envelope detectors for Gaussian noise. This comparison was made at a false alarm probability $p_f = 10^{-3}$. The detection loss (in dB) shown in Table 6.2 for different sample sizes n and number of quantizer levels M is the additional SNR S required by the QEC detector to obtain approximately the same performance as the linear and square-law envelope detectors in the non-fluctuating and Rayleigh fluctuating cases, respectively. In the latter case we see that in a typical situation with $n = 16$ pulses a four-level AO quantization gives an actual loss of about 1.0 dB when the asymptotic loss is 0.25 dB. Convergence to the asymptotic value is quite slow beyond $n = 32$, for which the loss is 0.6 dB. This level of loss is, nonetheless, quite small, considering that only two-bit quantization is employed. For non-fluctuating pulses we find that even the use of only 8 pulses results in a loss of only 0.4 dB for four-level quantization, compared to 0.25 dB asymptotically. It should be noted that for larger samples sizes (and smaller SNRs to obtain, say, a fixed detection probability of 0.5) the linear envelope detector is not a good approximation to the optimum detector; the LO square-law detector begins to become a better approximation. This is reflected in the fact that beyond $n = 32$ for four-level quantization (and beyond comparable values of n for fewer output quantization levels) the AO quantizer is *better* than the linear envelope detector. We have used the linear envelope detector for this comparison because it is widely used as an "optimum" detector for Gaussian noise and the non-fluctuating pulse amplitudes model.

	M	4	8	16	32	64	128	∞
NONFLUCTUATING TARGET	2	3.5	1.5	1.3	1.2	0.8	0.8	0.94
	3	1.5	0.8	0.5	0.5	0.5	0.4	0.43
	4	0.6	0.4	0.3	0.3	0.2	0.2	0.25
PULSE-TO-PULSE RAYLEIGH FLUCTUATING TARGET	2	5.0	4.0	2.5	1.7	1.6	1.5	0.94
	3	3.0	2.0	1.3	1.0	0.8	0.7	0.43
	4	2.0	1.5	1.0	0.6	0.5	0.5	0.25

Table 6.2 Finite-Sample-Size Detection Loss of QEC Detectors, in Gaussian Noise

We have given numerical finite-sample-size performance results only for the Gaussian noise case and constant per-pulse SNR ($e_i = 1$, all i). However, this comparison with asymptotic results indicates that the general quantizer design based on asymptotic performance criteria leads to systems which can be expected to perform reasonably close to their asymptotic predicted operating points in typical applications. It is, of course, possible to fine-tune the quantizer design for finite sample sizes through numerical performance evaluations [Hansen, 1974]. Any gains in performance obtained will generally remain small. One of the earliest investigations of multilevel quantization of the envelope detected outputs for the narrowband pulse-train detection problem was made by Hansen [1969]. More recently this problem has been addressed by Cimini and Kassam [1983].

6.4.3 M-Region Generalized Quantization

To complete our discussion of AO quantization in this section let us briefly consider AO generalized quantization as we did for the low-pass signal case in Section 4.3 of Chapter 4. We can also further generalize our result by dropping the assumption that U_{Ii} and U_{Qi} in (6-58) and (6-59) have a circularly symmetric density function. We shall do so now because this does not significantly complicate the analysis; in most cases the circular symmetry would, however, be a reasonable assumption.

Let $\{A_j, \ j = 1,2,...,M\}$ now be some partition of the two-dimensional space $\mathbb{R}^2$. Using (6-76) we can immediately obtain the normalized efficacy of an M-region quantizer using levels $l_1, l_2, \ldots, l_M$ on these subsets as

$$\tilde{\xi}_{QC} = \frac{1}{4} \frac{\left[\sum_{j=1}^{M} l_j \int_{A_j} g_{LO}(u,v) f_{IQ}(u,v) \, du \, dv \right]^2}{\sum_{j=1}^{M} l_j^2 \int_{A_j} f_{IQ}(u,v) \, du \, dv} \quad (6\text{-}102)$$

where g_{LO} is now the LO function

$$g_{LO}(u,v) = \frac{1}{2} \frac{\nabla^2 f_{IQ}(u,v)}{f_{IQ}(u,v)} \quad (6\text{-}103)$$

The LO levels l_j which maximize $\tilde{\xi}_{QC}$ may now be taken to be

$$l_j = \frac{\int_{A_j} g_{LO}(u,v) f_{IQ}(u,v) \, du \, dv}{\int_{A_j} f_{IQ}(u,v) \, du \, dv} \quad (6\text{-}104)$$

and with these levels the normalized efficacy becomes

$$\tilde{\xi}_{QC,LO} = \frac{1}{4} \sum_{j=1}^{M} \frac{[\int_{A_j} g_{LO}(u,v) f_{IQ}(u,v) \, du \, dv]^2}{\int_{A_j} f_{IQ}(u,v) \, du \, dv} \quad (6\text{-}105)$$

We can now proceed exactly as we did in Section 4.3 of Chapter 4 to obtain the AO subsets A_j in the M-region partitioning. Skipping the details now, which are very similar, we will state only the final result: the set of regions $\{A_j, \; j = 1,2,...,M\}$ maximizing $\tilde{\xi}_{QC,LO}$ of (6-105) necessarily satisfy

$$A_j = \{(u,v) \mid \tilde{t}_{j-1} < g_{LO}(u,v) \leq \tilde{t}_j\}, \; j = 1,2,...,M \quad (6\text{-}106)$$

for some set of constants $0 = \tilde{t}_0 \leq \tilde{t}_1 \leq \cdots \leq \tilde{t}_M = \infty$. In particular, the $\tilde{t}_j$ should be

$$\tilde{t}_j = \frac{l_j + l_{j+1}}{2}, \; j = 1,2,...,M-1 \quad (6\text{-}107)$$

with the l_j given by (6-104). Thus (6-104) and (6-106) must be solved simultaneously for the AO quantizer. In stating our result for the AO regions as (6-106) we have assumed that the events $g_{LO}(X_{Ii}, X_{Qi}) = \tilde{t}_j$ have zero probability. This quantization problem has also been treated by Cimini and Kassam [1983].

The result (6-106) shows that if $g_{LO}(u,v) = \tilde{p}_{LO}(\sqrt{u^2 + v^2})$, which is the case for a circularly symmetric bivariate density function, then the AO generalized quantization scheme does reduce to an AO generalized *envelope* quantization scheme. If $\tilde{p}_{LO}$ is in addition a monotone function, then the M-region envelope quantizer does reduce to an M-interval envelope quantizer.

Our analyses so far of quantized data detection systems have given us very similar general results for the different situations we have considered (low-pass signal, coherent, and incoherent narrowband signal, noncoherent pulse train). We shall obtain another basically similar set of results for the random signal detection problems we will consider in the next chapter.

PROBLEMS

Problem 6.1

In the observation model given by (6-4) and (6-5) replace θ by θA_i, the A_i being i.i.d. random amplitudes independent of ψ with a common pdf f_A. Derive the test statistic of the LO detector for $\theta = 0$ versus $\theta \neq 0$ in this "fast-fading" situation.

Problem 6.2

Show that the ARE of two quadrature GNC/envelope detectors using characteristics $\tilde{g}_1$ and $\tilde{g}_2$ for circularly symmetric bivariate noise pdf in testing $\theta = 0$ versus $\theta \neq 0$ in (6-4) and (6-5) is the same as the corresponding ARE for the coherent narrowband signal detection problem. (Use conditional expectations, with conditioning on ψ.)

Problem 6.3

A special case of the test statistic of (6-49) is obtained with $R_i \tilde{g}(R_i)$ a constant. The resulting test statistic is equivalent to the quadrature hard-limiter narrowband correlator statistic

$$T_{HNC}(\tilde{\mathbf{X}}) = \left| \sum_{i=1}^{n} \tilde{s}_i \tilde{U}_i^* \right|^2$$

where $\tilde{U}_i = \tilde{X}_i / R_i$ and $R_i = |\tilde{X}_i|$.

Show that in testing $\theta = 0$ versus $\theta \neq 0$ in the model (6-4) and (6-5) when f_{IQ} is circularly symmetric, any desired false-alarm probability value can be maintained with a fixed threshold for all noise-only circularly symmetric pdf's.

Problem 6.4

In $T_{HNC}(\tilde{\mathbf{X}})$ of Problem 6.3, consider a uniform phase quantization of the unit amplitude $\tilde{U}_i$ into M values equally spaced on the unit circle. Obtain the efficacy of the resulting detector for testing $\theta = 0$ versus $\theta \neq 0$ in the model (6-4) and (6-5) with circularly symmetric bivariate noise pdf f_{IQ}.

Problem 6.5

Let $\tilde{g}_a(r)$ be the limiter function

$$\tilde{g}_a(r) = \begin{cases} r/a, & 0 \leq r \leq a \\ 1, & r > a \end{cases}$$

(a) Obtain the efficacy of a generalized envelope correlator detector using the statistic

$$T_{GEC}(\mathbf{R}) = \sum_{i=1}^{n} e_i^z \, \tilde{g}(R_i)$$

with $\tilde{g}(R_i) = \tilde{g}_a(R_i)$, for Gaussian noise in the noncoherent pulse train detection problem. Compare the maximum efficacy for optimum a with the efficacy of the AO square-law envelope correlator in this situation.

(b) Repeat the above for the circularly symmetric noise pdf for which $h(r)$ is given by (6-67).

Problem 6.6

For the noncoherent pulse train detection problem consider the dead-zone limiter function

$$g_A(u,v) = \begin{cases} 1, & (u,v) \in A \\ 0, & (u,v) \in \overline{A} \end{cases}$$

for use as g in (6-71). Show that the AO region A maximizing the efficacy also results in minimum mean-squared error under the null hypothesis between $\nabla^2 f_{IQ}(u,v)/f_{IQ}(u,v)$ and the two-level function with best choice of levels l_1 and l_2 and of corresponding regions $\overline{A}$ and A. Obtain explicitly the simultaneous equations which have to be solved for the AO region A.

Chapter 7

DETECTION OF RANDOM SIGNALS IN ADDITIVE NOISE

7.1 Introduction

Having discussed the detection of known signals in additive white noise and the detection of narrowband coherent and incoherent signals in narrowband noise, we finally turn our attention to the problem of random signal detection. In the detection of narrowband incoherent signals we allowed the phase (as well as the amplitude) of the signal to be modeled as a random variable. Here our signal will be modeled entirely as a non-deterministic random process. Following the scheme of our development so far, the focus will be on the detection of a random signal in additive white noise. Additionally, however, a brief discussion of narrowband random signal detection in narrowband noise is included in this chapter.

In many applications involving detection of random signals, the observations are obtained simultaneously from a number of receivers forming an *array*. The detection problem then often becomes that of detecting a random signal which is *common* to each receiving element, embedded in noise processes at the receiving elements, which are uncorrelated with each other and with the signal process. A typical example of this type of application is an underwater hydrophone array used in a passive sonar system to detect the presence of, and to locate, sources of random signals such as submarines. In operation the array searches for sources by imparting relative time delays to the outputs of the receiving elements, to equalize the propagation delays to each receiver for signals emanating from a search direction. This is followed by a detector for a common random signal.

With a single receiving element the bearing of the source of any detected signal cannot be established. In fact, it is often extremely difficult to even detect the presence of a weak random signal source with only a single receiver. Suppose, for example, that the classical assumption of Gaussian processes is made for the uncorrelated signal and noise. If both are stationary white processes (or if both have the same normalized autocovariance function) and if the noise power level is not known, it is clearly impossible to detect the presence of a signal. But in an array of receivers with uncorrelated noise the presence of a common signal causes the individual outputs to be positively correlated, in addition to increasing the received power at each receiver. Thus signal detectability is greatly enhanced. In this section we will treat the problem of common random signal detection in an array of receivers. The case of a single receiver will then obviously be included as a special case.

Once again we will consider *locally optimum* schemes for weak-signal detection. For random signal detection problems the statistics of the signal process in addition to those of the noise process are involved in any derivation of *optimum* detection schemes. Thus it is rare to find that a UMP detector exists. As we shall see, the locally optimum detector structure turns out to depend only on the mean and covariance functions of the signal process, and does not require knowledge of density functions characterizing the signal process, which may be non-Gaussian.

Beyond our development of the locally optimum detector structure in Section 7.3, we give relatively brief accounts of asymptotic performance, of optimum quantization, and of narrowband random signals in the rest of the chapter. In addition, we have not included a discussion of the locally optimum Bayes scheme for random signal detection. These topics may be given detailed treatment in much the same way as we have done previously for other versions of the signal detection problem.

7.2 The Observation Model

Let $\mathbf{X}_j = (X_{j1}, X_{j2}, ..., X_{jn})$ be the vector of n observation components obtained at the j-th element of a receiving array of L elements. As in our other observation models, X_{ji} may be thought of as the i-th time-sampled value of a continuous-time observation process at the j-th array element. More generally, it could be the sampled output of some pre-processor in the j-th receiver channel. At each time index i the observations X_{ji}, $j = 1,2,...,L$ across the array may have a common random signal S_i in additive noise components W_{ji}. Our observation model is thus

$$X_{ji} = \theta S_i + W_{ji}, \quad \begin{cases} j = 1,2,...,L \\ i = 1,2,...,n \end{cases} \qquad (7\text{-}1)$$

Let us now list the basic assumptions we will make about the signal and noise quantities in our model, and also introduce the notation we will use.

1) The noise components W_{ji} are i.i.d. random variables with a known common probability density function f for $j = 1,2,...,L$ and $i = 1,2,...,n$. The noise components have zero means and variances $\sigma^2 = 1$.

2) The signal process $\{S_i\}_{i=1}^{n}$ has a known mean sequence $\{\mu_i\}_{i=1}^{n}$. The centered process $\{S_i - \mu_i\}_{i=1}^{n}$ is stationary and has a variance $\sigma_S^2 = 1$. The known autocovariance function of the signal process is $r_S(i)$, $i = 0, \pm 1, \ldots$. The univariate probability density function of $S_i - \mu_i$ is f_S, and the probability density function of the centered signal vector $\mathbf{S} - \boldsymbol{\mu} = (S_1 - \mu_1, S_2 - \mu_2, \ldots, S_n - \mu_n)$ is $f_{\mathbf{S}}$.

3) The signal and noise processes are independent.

The assumption that the signal and noise variances are unity is not restrictive since the parameter θ controls the SNR, which is now simply θ^2. The data may, for example, be assumed to have been normalized by σ. While we have imposed some restrictions in our assumptions above, notably that of independence of the W_{ji}, the model is general enough to allow some interesting results to be obtained. The temporal independence of the noise components is essential to allow non-Gaussian noise to be considered with any degree of success in terms of obtaining explicit solutions for useful detection schemes. While it is possible to consider correlated noise *across* the array at each time instant, since at most L-variate non-Gaussian densities are involved, we will not do so here. The more restrictive assumption of independence is often reasonable and does allow considerable simplification of the analysis and the practical interpretation of the results.

Although we will begin by allowing the mean vector $\boldsymbol{\mu} = (\mu_1, \mu_2, \ldots, \mu_n)$ to be non-zero, most of the results which are of special interest are obtained for the zero-mean signal. This is because when $\boldsymbol{\mu}$ is not the zero vector the results are essentially those for detection of a known deterministic signal.

Let $\mathbf{X}$ now denote the $L \times n$ matrix of row vectors $\mathbf{X}_j$, $j = 1, 2, \ldots, L$ and let $f_{\mathbf{X}}$ be the joint probability density function of the Ln components of $\mathbf{X}$. The random signal detection problem can then be stated as a problem of testing a null hypothesis H_3 versus an alternative hypothesis K_3 about the density function $f_{\mathbf{X}}$:

$$H_3: f_{\mathbf{X}}(\mathbf{x}) = \prod_{i=1}^{n} \prod_{j=1}^{L} f(x_{ji}) \qquad (7\text{-}2)$$

$$K_3: f_{\mathbf{X}}(\mathbf{x}) = E\left\{\prod_{i=1}^{n} \prod_{j=1}^{L} f(x_{ji} - \theta S_i)\right\}, \text{ any } \theta \neq 0 \qquad (7\text{-}3)$$

The expectation in (7-3) is, of course, taken with respect to $\mathbf{S}$ and under K_3 $f_{\mathbf{X}}$ may be expressed explicitly as

$$f_{\mathbf{X}}(\mathbf{x} \mid \theta) = \int_{\mathbf{R}^n} \prod_{i=1}^{n} \prod_{j=1}^{L} f(x_{ji} - \theta s_i) f_{\mathbf{S}}(\mathbf{s} - \boldsymbol{\mu}) d\mathbf{s}, \quad \theta \neq 0 \qquad (7\text{-}4)$$

Throughout the remainder of this chapter some regularity conditions are necessary on f to make the derivations valid. To avoid detailed mathematical considerations we do not specify precisely these conditions, but assume simply that f is regular enough to allow the results to be valid.

There is another hypothesis-testing problem, called the *scale-change* problem, which may also be viewed as a formulation of a random signal detection problem. Consider, for instance, the detection of a Gaussian signal in independent additive Gaussian noise. Then the univariate density function of the observations is always Gaussian, with a larger value for the *variance* under the alternative hypothesis. Generalization of this Gaussian situation to non-Gaussian densities gives us a hypothesis-testing problem where the density function of the observations is of some known type, with different values for a scale parameter under the null and alternative hypotheses. The LO test for this problem has been investigated by Poor and Thomas [1978b]; we shall not deal with this model here, since in the non-Gaussian case it is not generally appropriate for *additive* signals in noise.

7.3 Locally Optimum Array Detection

In this section we will first obtain the general form of the LO detector for our problem, using approaches which we have by this point already applied at several places in earlier chapters. We then consider the special case of a white signal sequence, and finally show that for a non-white zero-mean signal sequence the LO detector utilizes the Eckart filter in forming its test statistic.

7.3.1 General Form of the LO Test Statistic

The derivation of the LO detector is facilitated with the following definitions.

$$P(\theta) \triangleq \prod_{i=1}^{n} \prod_{j=1}^{L} f(x_{ji} - \theta s_i) \tag{7-5}$$

$$P_{lk}(\theta) \triangleq P(\theta)/f(x_{lk} - \theta s_k) \tag{7-6}$$

and

$$P_{lk,pm}(\theta) \triangleq P_{lk}(\theta)/f(x_{pm} - \theta s_m) \tag{7-7}$$

Now to obtain the test statistic of the LO detector, we can differentiate f_X of (7-4) with respect to the signal strength parameter θ, and evaluate it at $\theta = 0$.

Writing $f_X(\mathbf{x})$ under K_3 as

$$f_X(\mathbf{x} \mid \theta) = \int_{\mathbf{R}^n} P(\theta) f_S(\mathbf{s} - \boldsymbol{\mu}) \, d\mathbf{s} \tag{7-8}$$

the derivative with respect to θ becomes

$$\frac{d}{d\theta} f_X(\mathbf{x} \mid \theta) = \int_{\mathbf{R}^n} \frac{dP(\theta)}{d\theta} f_S(\mathbf{s} - \boldsymbol{\mu}) \, d\mathbf{s}$$

$$= -\int_{\mathbf{R}^n} \sum_{i=1}^{n} \sum_{j=1}^{L} P_{ji}(\theta) f'(x_{ji} - \theta s_i) s_i f_S(\mathbf{s} - \boldsymbol{\mu}) \, d\mathbf{s} \tag{7-9}$$

and

$$\left. \frac{d}{d\theta} f_X(\mathbf{x} \mid \theta) \right|_{\theta=0} = -P(0) \sum_{i=1}^{n} \sum_{j=1}^{L} \mu_i \frac{f'(x_{ji})}{f(x_{ji})} \tag{7-10}$$

Then the LO test statistic becomes

$$T_{LO}(\mathbf{X}) = \frac{\left. \frac{d}{d\theta} f_X(\mathbf{X} \mid \theta) \right|_{\theta=0}}{\left. f_X(\mathbf{X} \mid \theta) \right|_{\theta=0}}$$

$$= \sum_{i=1}^{n} \mu_i \sum_{j=1}^{L} \frac{-f'(X_{ji})}{f(X_{ji})} \tag{7-11}$$

This LO statistic is simply an extension to the array case of our result for known *deterministic* signal detection of Chapter 2, and has been derived without utilizing any second-order information about the signal process.

If the mean values μ_i are *all* equal to zero then $\frac{d}{d\theta} f_X(\mathbf{x} \mid \theta)$ evaluated at $\theta = 0$ is zero, and the power functions of all tests for our problem have zero slope at $\theta = 0$. In this case we have to maximize the second derivative of the power function at $\theta = 0$. As we have seen in Section 1.4 of Chapter 1, this means that the LO statistic now requires evaluation of $\frac{d^2}{d\theta^2} f_X(\mathbf{x} \mid \theta)$ at $\theta = 0$. From (7-9) we get

$$\frac{d^2}{d\theta^2} f_X(\mathbf{x} \mid \theta) = \sum_{i=1}^{n} \sum_{j=1}^{L} \int_{\mathbf{R}^n} -\frac{d}{d\theta} [P_{ji}(\theta) f'(x_{ji} - \theta s_i)] s_i f_S(\mathbf{s} - \boldsymbol{\mu}) \, d\mathbf{s}$$

$$= \sum_{i=1}^{n} \sum_{j=1}^{L} \int_{\mathbf{R}^n} P_{ji}(\theta) f''(x_{ji} - \theta s_i) s_i^2 f_S(\mathbf{s} - \boldsymbol{\mu}) \, d\mathbf{s}$$

$$+ \sum_{i=1}^{n} \sum_{j=1}^{L} \int_{\mathbb{R}^n} -f'(x_{ji}-\theta s_i) \frac{d}{d\theta} P_{ji}(\theta) s_i \, f\, \mathbf{s}(\mathbf{s}-\boldsymbol{\mu}) \, d\mathbf{s}$$

(7-12)

Now

$$\frac{d}{d\theta} P_{ji}(\theta) = - \sum_{\substack{m=1 \\ (p,m) \neq (j,i)}}^{n} \sum_{p=1}^{L} P_{ji,pm}(\theta) f'(x_{pm}-\theta s_m) s_m \quad (7\text{-}13)$$

so finally

$$\frac{d^2}{d\theta^2} f_{\mathbf{x}}(\mathbf{x}\mid\theta)\bigg|_{\theta=0} = P(0) \sum_{i=1}^{n} \sum_{j=1}^{L} (1+\mu_i^2) \frac{f''(x_{ji})}{f(x_{ji})}$$
$$+ P(0) \sum_{i=1}^{n} \sum_{j=1}^{L} \sum_{\substack{m=1 \\ (p,m)\neq(j,i)}}^{n} \sum_{p=1}^{L} E\{S_i S_m\} \frac{f'(x_{ji})}{f(x_{ji})} \frac{f'(x_{pm})}{f(x_{pm})}$$

(7-14)

We then obtain

$$\frac{d^2}{d\theta^2} f_{\mathbf{x}}(\mathbf{X}\mid\theta)\bigg|_{\theta=0} = P(0) \sum_{i=1}^{n} \sum_{j=1}^{L} (1+\mu_i^2) h_{LO}(X_{ji})$$
$$+ P(0) \sum_{i=1}^{n} \sum_{j=1}^{L} \sum_{m=1}^{n} \sum_{p=1}^{L} (r_S(m-i)+\mu_i \mu_m) g_{LO}(X_{ji}) g_{LO}(X_{pm})$$
$$= P(0) \sum_{i=1}^{n} (1+\mu_i^2) \sum_{j=1}^{L} h_{LO}(X_{ji})$$
$$+ P(0) \left[\sum_{i=1}^{n} \mu_i \sum_{j=1}^{L} g_{LO}(X_{ji})\right]^2$$
$$+ P(0) \sum_{i=1}^{n} \sum_{m=1}^{n} r_S(m-i) \sum_{j=1}^{L} \sum_{p=1}^{L} g_{LO}(X_{ji}) g_{LO}(X_{pm})$$

(7-15)

where

$$h_{LO}(x) \triangleq \frac{f''(x)}{f(x)} - g_{LO}^2(x)$$
$$= - g_{LO}'(x) \quad (7\text{-}16)$$

and, as before,

$$g_{LO}(x) = \frac{-f'(x)}{f(x)} \quad (7\text{-}17)$$

For *zero-mean* signal the LO test statistic is therefore

$$\lambda_{LO}(\mathbf{X}) = \sum_{i=1}^{n} \sum_{j=1}^{L} h_{LO}(X_{ji})$$
$$+ \sum_{i=1}^{n} \sum_{m=1}^{n} r_S(m-i) \sum_{j=1}^{L} g_{LO}(X_{ji}) \sum_{p=1}^{L} g_{LO}(X_{pm}) \quad (7\text{-}18)$$

This result depends on the signal statistics only through the signal autocovariance function.

Before we continue to discuss the above result for the LO test statistic, it is instructive to use our results so far to consider the second-order approximation of the likelihood ratio,

$$L(\mathbf{X}) = f_X(\mathbf{X} \mid \theta)/f_X(\mathbf{X} \mid 0)$$
$$\approx 1 + \theta \sum_{i=1}^{n} \mu_i \sum_{j=1}^{L} g_{LO}(X_{ji})$$
$$+ \frac{\theta^2}{2} \left\{ \left[\sum_{i=1}^{n} \mu_i \sum_{j=1}^{L} g_{LO}(X_{ji}) \right]^2 \right.$$
$$+ \sum_{i=1}^{n} (1 + \mu_i^2) \sum_{j=1}^{L} h_{LO}(X_{ji})$$
$$\left. + \sum_{i=1}^{n} \sum_{m=1}^{n} r_S(m-i) \sum_{j=1}^{L} g_{LO}(X_{ji}) \sum_{p=1}^{L} g_{LO}(X_{pm}) \right\}$$
$$(7\text{-}19)$$

When all the μ_i are not equal to zero, the statistic $T_{LO}(\mathbf{X})$ of (7-11) is the term retained in (7-19) for a first-order approximation to $L(\mathbf{X})$. If this term is zero, because $\mu_i = 0$ for all i, the most significant data-dependent term in the above approximation is $\theta^2 \lambda_{LO}(\mathbf{X})/2$. The factor of 2 can be explained by treating θ^2 as the signal strength parameter in this case; we have

$$\frac{d}{d(\theta)^2} f_X(\mathbf{X} \mid \theta) = \frac{1}{2\theta} \frac{d}{d\theta} f_X(\mathbf{X} \mid \theta) \quad (7\text{-}20)$$

and

$$\lim_{\theta^2 \to 0} \frac{d}{d(\theta^2)} f_X(\mathbf{X} \mid \theta) = \frac{1}{2} \lim_{\theta \to 0} \frac{d^2}{d\theta^2} f_X(\mathbf{X} \mid \theta) \quad (7\text{-}21)$$

when $\lim_{\theta \to 0} \frac{d}{d\theta} f_X(\mathbf{X} \mid \theta) = 0$. This expansion also allows us to note a distinction between AO and LO schemes. Suppose that all the μ_i are not zero, but we have $\lim_{n \to \infty} \sum_{i=1}^{n} \mu_i^2/n = 0$. Then $T_{LO}(\mathbf{X})$ will not produce an AO sequence of detectors (for $\theta = \gamma/\sqrt{n}$) but $\lambda_{LO}(\mathbf{X})$ will give an AO sequence of detectors for $\theta^2 = \gamma/\sqrt{n}$ under some regularity conditions for r_S.

7.3.2 LO Statistic for White Signal Sequence

Henceforth the assumption will be made that the signal process has *zero mean* so that $\lambda_{LO}(\mathbf{X})$ of (7-18) is the LO (and AO) test statistic. If the additional assumption is made that the signal is "white," so the $r_S(i) = 0$ for $i \neq 0$, the LO statistic simplifies to

$$\lambda_{LO}(\mathbf{X}) = \sum_{i=1}^{n} \left\{ \sum_{j=1}^{L} h_{LO}(X_{ji}) + \left[\sum_{j=1}^{L} g_{LO}(X_{ji}) \right]^2 \right\} \quad (7\text{-}22)$$

The structure of this LO test statistic is illustrated in Figure 7.1.

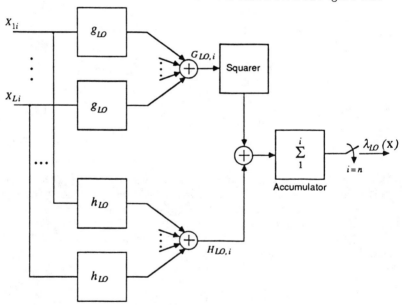

Figure 7.1 Locally Optimum Test Statistic for Random, White Signal in Additive Noise

An alternative expression for (7-22) is

$$\lambda_{LO}(\mathbf{X}) = \sum_{j=1}^{L} \sum_{i=1}^{n} \frac{f''(X_{ji})}{f(X_{ji})}$$

$$+ 2 \sum_{j=1}^{L} \sum_{p=j+1}^{L} \sum_{i=1}^{n} g_{LO}(X_{ji}) g_{LO}(X_{pi}) \qquad (7\text{-}23)$$

In this form $\lambda_{LO}(\mathbf{X})$ can be interpreted as being composed of two terms: one is the sum of GC statistics formed for each channel, and the second is a sum over distinct pairs of channels of a *generalized cross-correlation* (GCC) statistic. The structure of (7-23) shows clearly the two distinct features which the detector attempts to use in detecting the random signal. The first term in $\lambda_{LO}(\mathbf{X})$ above is a generalized measure of the *energy* received by the array. To appreciate this better, assume that f is a symmetric function. Then f''/f is also an even function. For example, for unit-variance Gaussian noise we have

$$\frac{f''(x)}{f(x)} = x^2 - 1 \qquad (7\text{-}24)$$

The second term in (7-23) is a generalized measure of the *correlation* between observations in different channels of the array. This structure for the LO test statistic is reasonable from intuitive considerations, since the presence of a common random signal both increases the power level of the received data and causes data in different receiver channels to become correlated.

It is important to note that the increase in power level occurs whenever random signals are present at the individual receivers of the array regardless of whether the signals *across* the array are one common signal or are completely uncorrelated. The GCC part of the LO statistic responds only to a *common* signal or at least to signals which are spatially correlated across the array elements. This is a major reason why it is useful to employ only the GCC part of the LO statistic in applications involving detection as well as location of signal sources.

Of course when only one receiver is present ($L = 1$) the system can only act as a signal detector. In this case the LO test statistic of (7-23) reduces to

$$\lambda_{LO}(\mathbf{X}_1) = \sum_{i=1}^{n} e_{LO}(X_{1i}) \qquad (7\text{-}25)$$

where

$$\begin{aligned} e_{LO}(x) &= h_{LO}(x) + g_{LO}^2(x) \\ &= \frac{f''(x)}{f(x)} \end{aligned} \qquad (7\text{-}26)$$

As noted above, this is a *generalized energy* (GE) detector, a special case of which is the *square-law* detector, which is LO for Gaussian noise. The square-law detector is in fact the UMP detector for a *Gaussian* white signal in Gaussian white noise.

It is not possible to apply the above result (for $L=1$) directly to obtain the LO detector when noise has the *double-exponential* density function, such a probability density function not being regular enough at the origin. The unit variance double-exponential density function is

$$f(x) = \frac{1}{\sqrt{2}} e^{-\sqrt{2}|x|} \qquad (7\text{-}27)$$

with

$$f'(x) = -\sqrt{2}\,\text{sgn}(x)f(x) \qquad (7\text{-}28)$$

so that

$$f''(x) = 2f(x) - 2\sqrt{2}\,\delta(x)f(x) \qquad (7\text{-}29)$$

where δ is the Dirac delta function. Thus a *formal* use of our result would imply that the LO statistic is now a GE statistic based on use of the function

$$\begin{aligned} e_{LO}(x) &= \frac{f''(x)}{f(x)} \\ &= 2 - 2\sqrt{2}\,\delta(x) \end{aligned} \qquad (7\text{-}30)$$

This result suggests that a possible good nonlinearity for use in the GE detector for such noise is the "dead-zone" function

$$d_\epsilon(x) = \begin{cases} 1, & |x| > \epsilon \\ 0, & |x| \leq \epsilon \end{cases} \qquad (7\text{-}31)$$

for some small constant $\epsilon > 0$. To arrive at the above a narrow-pulse approximation to $\delta(x)$ is used, together with amplitude-scaling and addition of a constant to the result of (7-30). Notice that with the above function in the GE test statistic

$$T_{GE}(\mathbf{X}_1) = \sum_{i=1}^{n} e(X_{1i}) \qquad (7\text{-}32)$$

the test statistic has a binomial distribution under both hypotheses. The binomial parameter $p = P\{|X_{1i}| > \epsilon\}$ is larger under the alternative, signal-present hypothesis. It is possible to analyze the finite sample-size performance of this "dead-zone" GE detector and obtain the optimum value of ϵ for given n and θ. We shall see in the next section that the *efficacy* of such a detector when the noise has a double-exponential density function grows without bound as $\epsilon \to 0$. This behavior for the locally optimum detector for double-exponential noise density is in general accord with our result for the known signal case, in which the sign detector was obtained. We have noted earlier that for such a noise density function the magnitudes of the observations are not important for weak deterministic signal detection.

Consider now the second term in (7-23), for $L > 1$, as a sub-optimum statistic

$$\hat{\lambda}_{LO}(\mathbf{X}) = 2 \sum_{j=1}^{L} \sum_{p=j+1}^{L} \left[\sum_{i=1}^{n} g_{LO}(X_{ji}) \, g_{LO}(X_{pi}) \right] \quad (7\text{-}33)$$

As has been remarked above, this is a sum of generalized cross-correlation statistics for distinct pairs of channels. A GCC statistic for the j-th and p-th channels may be defined as

$$T_{GCC}(\mathbf{X}_j, \mathbf{X}_p) = \sum_{i=1}^{n} g(X_{ji}) \, g(X_{pi}) \quad (7\text{-}34)$$

where g is the characteristic nonlinearity of the GCC with $E\{g(W_{ji})\} = 0$. For g the linear function, for example, we get the *linear cross-correlation* (LCC) statistic. We will see in the next section that the choice $g = g_{LO}$ gives us the *asymptotically optimum* GCC test statistic; of course, we know that from among the class of *all* possible detectors (not constrained to GCC detectors) this only gives us a sub-optimum scheme. The use of GCC statistics is attractive because the presence of a signal is now sensed by the increase in the correlation (from zero) between $g(X_{ji})$ and $g(X_{pi})$. To set the detection threshold to achieve some design value for the false-alarm probability one may use a simple technique to achieve the requisite performance at least asymptotically, for large n. Specifically, the statistic $T_{GCC}(\mathbf{X}_j, \mathbf{X}_p)$ may be normalized by an estimate of the variance of $g(X_{ji})$. This is done using the observed data, regardless of whether the signal is known to be absent or not. Then under the null hypothesis the asymptotic distribution of the normalized GCC statistic may be determined. A special case of this is the LCC statistic, for which the normalization is by an estimate of the variance of the observations, and the normalized statistic is simply the coefficient of correlation.

We note in passing that with $g(x) = \text{sgn}(x)$ in (7-34) we get the *polarity coincidence correlator* (PCC), a special form of the GCC. This is an AO GCC detector for noise with the double-exponential density function (as we will see), and is an example of a *nonparametric* detector [Kassam and Thomas, 1980].

7.3.3 LO Statistic for Correlated Signal Sequence

Let us return now to the general form of the LO statistic for a zero-mean signal which has correlated components S_i. Upon defining the processed array quantities (as defined in Figure 7.1)

$$H_{LO,i} \triangleq \sum_{j=1}^{L} h_{LO}(X_{ji}) \qquad (7\text{-}35)$$

and

$$G_{LO,i} \triangleq \sum_{j=1}^{L} g_{LO}(X_{ji}) \qquad (7\text{-}36)$$

for all i, with these quantities taken to be zero when i is outside the range $1 \leq i \leq n$, we get from (7-18)

$$\lambda_{LO}(\mathbf{X}) = \sum_{i=-\infty}^{\infty} H_{LO,i} + \sum_{i=-\infty}^{\infty} \sum_{k=-\infty}^{\infty} r_S(k) \, G_{LO,i} \, G_{LO,i+k} \qquad (7\text{-}37)$$

Now since the sequence r_S is non-negative definite, we can in general express it as

$$r_S(k) = \sum_{l=-\infty}^{\infty} a_{k+l} \, a_l^* \qquad (7\text{-}38)$$

with $a_l = 0$ for $l < 0$; here the sequence $\{a_k\}_{k=0}^{\infty}$ is the impulse response of a filter with frequency response $H(\omega)$ satisfying

$$|H(\omega)|^2 = \left| \sum_{l=0}^{\infty} a_l \, e^{-jl\omega} \right|^2$$

$$= \Phi_S(\omega) \qquad (7\text{-}39)$$

where $\Phi_S(\omega)$ is the signal power spectral density. In particular, for signal spectrum $\Phi_S(\omega)$ satisfying the Paley-Wiener criterion, a unique representation (7-38) is obtained when $\{a_k\}_{k=0}^{\infty}$ is a minimum-phase sequence. From (7-38) and (7-37) we can obtain the result

$$\lambda_{LO}(\mathbf{X}) = \sum_{i=1}^{n} H_{LO,i} + \sum_{i=-\infty}^{\infty} \left| \sum_{k=1}^{n} a_{i-k} \, G_{LO,k} \right|^2 \qquad (7\text{-}40)$$

Note that the quantities under the magnitude-squared operation above are the outputs of the filter $H(\omega)$ with inputs the $G_{LO,i}$.

If the coherence time of the signal is short compared to its duration n (large time-bandwidth product for the signal) and the impulse-response sequence $\{a_k\}_{k=-\infty}^{\infty}$ is negligible outside some interval around the origin which is short compared to n, the infinite sum in (7-40) may be replaced by a summation over the range $1 \leq i \leq n$. This approximation results in the implementation of the LO detector shown in Figure 7.2.

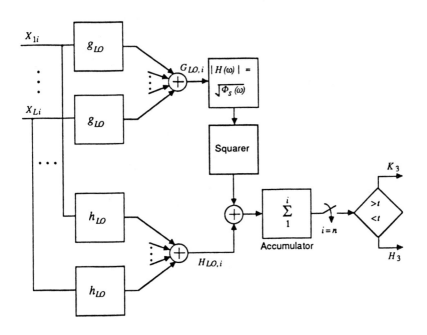

Figure 7.2 Locally Optimum Detector for Random Signal in Additive Noise

If the noise is not white, then our result for the LO statistic is still correct if the input data in each channel are first "whitened" by passage through a filter with frequency response $1/\sqrt{\Phi_W(\omega)}$, where $\Phi_W(\omega)$ is the noise spectrum. Note that in this case f is the univariate probability density function of the resulting white noise, and also that r_S is now the autocovariance function corresponding to the spectrum $\Phi_S(\omega)/\Phi_W(\omega)$. This also means that the filter in Figure 7.2 will now have an amplitude frequency response of $\sqrt{\Phi_S(\omega)}/\sqrt{\Phi_W(\omega)}$. For *Gaussian* noise for which $G_{LO,i}$ is a linear function of the X_{ji}, $j = 1,2,...,L$, the whitening filter and the second filter may be combined as one input filter with frequency response $\sqrt{\Phi_S(\omega)}/\Phi_W(\omega)$, which is the classical *Eckart filter* (Problem 7.4). The filter with frequency response $\sqrt{\Phi_S(\omega)}$ in Figure 7.2 is the Eckart filter for the white-noise case.

The locally optimum and asymptotically optimum detection of random signals in non-Gaussian noise has been considered in the early work of Middleton [1960, 1966] and Rudnick [1961]. More recent investigations of this problem have been made by Sheehy [1978] and Poor and Thomas [1978b]. Other results include those of Kassam [1978], Dwyer [1980], and Lu and Eisenstein [1981]. An interesting interpretation of the LO detector as an LO estimator-correlator structure has been given by Gardner [1982]. Schwartz and Vastola [1983] have considered impulsive interference modeled by the Middleton class A model in detection of random signals. Detection of random signals in dependent noise samples has been considered by Halverson and Wise [1981, 1984a, 1984b].

7.4 Asymptotic Performance Characteristics

We shall now obtain expressions for the efficacies of some of the test statistics which we have examined above. The generalized energy detector, the generalized cross-correlation array detector, as well as a generalization of the LO detector statistic for uncorrelated signal terms will be considered. We will thus concentrate on test statistics which are suggested by our results for LO detection of an uncorrelated signal sequence. The efficacy expressions are valid for the general hypothesis-testing problem H_3 versus K_3. Of course, their use in ARE studies requires asymptotic normality to hold, and this is most easily shown for uncorrelated signal sequences or for signal sequences having a finite correlation time.

Generalized Energy Detector ($L = 1$)

For the case $L = 1$, we will first derive the efficacy of the generalized energy detector based on the statistic $T_{GE}(\mathbf{X}_1)$ of (7-32). As is the case for the LO characteristic $e_{LO} \equiv f''/f$, we will additionally assume that $E\{e(X_{1i}) \mid H_3\} = 0$. Furthermore, it will be assumed that e, f, and f_S satisfy the general conditions needed to make the following analysis valid. We get

$$E\{e(X_{ji}) \mid K_3\} = \int_{-\infty}^{\infty} e(x) \int_{-\infty}^{\infty} f(x - \theta s) f_S(s) \, ds \, dx \quad (7\text{-}41)$$

and

$$\frac{d}{d(\theta)^2} E\{e(X_{ji}) \mid K_3\} = \int_{-\infty}^{\infty} e(x) \frac{1}{2\theta} \int_{-\infty}^{\infty} -sf'(x - \theta s) f_S(s) \, ds \, dx \quad (7\text{-}42)$$

Using the fact that $E\{S_i\} = 0$ and $E\{S_i^2\} = 1$, we obtain

$$\frac{d}{d(\theta)^2} E\{e(X_{ji}) \mid K_3\} \bigg|_{\theta^2 = 0} = \frac{1}{2} \int_{-\infty}^{\infty} e(x) f''(x) \, dx \quad (7\text{-}43)$$

This gives for the efficacy of the generalized energy detector the result

$$\xi_{GE} = \frac{1}{4} \frac{\left[\int_{-\infty}^{\infty} e(x) f''(x) \, dx \right]^2}{\int_{-\infty}^{\infty} e^2(x) f(x) \, dx} \quad (7\text{-}44)$$

Notice that this efficacy is maximized when e is the LO function of (7-26).

For the special case when the noise is Gaussian the LO detector is a square-law detector using the test statistic

$$T_{SQ}(\mathbf{X}_1) = \sum_{i=1}^{n} [X_{1i}^2 - 1] \quad (7\text{-}45)$$

With $e(x) = x^2 - 1$ in (7-44) we get the efficacy of the square-law detector as

$$\xi_{SQ} = \frac{1}{4} \frac{\left[\int_{-\infty}^{\infty} (x^2 - 1) f''(x) \, dx \right]^2}{\int_{-\infty}^{\infty} (x^4 - 2x^2 + 1) f(x) \, dx}$$

$$= \frac{1}{\int_{-\infty}^{\infty} x^4 f(x) \, dx - 1} \quad (7\text{-}46)$$

where f is a zero-mean unit-variance noise density function. The efficacy depends on the fourth moment of the noise density function, which has the value 3 for unit-variance Gaussian noise. Thus for Gaussian noise

$$\xi_{SQ} = 0.5 \tag{7-47}$$

For heavier-tailed unit-variance noise densities we can expect the fourth moment to be larger. For example, for unit-variance double-exponential noise density this fourth moment has the value 6, giving only

$$\xi_{SQ} = 0.2 \tag{7-48}$$

Consider now the *dead-zone generalized energy* (DZE) detector based on the function d_ϵ of (7-31), for a symmetric noise probability density function f. It is quite easy to obtain its efficacy directly, from

$$E\{d_\epsilon(X_{1i}) \mid K_3\} = 1 - \int_{-\epsilon}^{\epsilon} \int_{-\infty}^{\infty} f(x - \theta s) f_S(s) \, ds \, dx$$

$$= 1 - \int_{-\infty}^{\infty} [F(\epsilon - \theta s) - F(-\epsilon - \theta s)] f_S(s) \, ds \tag{7-49}$$

and

$$V\{d_\epsilon(X_{1i}) \mid H_3\} = 2[1 - F(\epsilon)] - 4[1 - F(\epsilon)]^2$$

$$= 2[1 - F(\epsilon)][2F(\epsilon) - 1] \tag{7-50}$$

From (7-49) it follows that

$$\frac{d}{d(\theta)^2} E\{d_\epsilon(X_{1i}) \mid K_3\} \bigg|_{\theta^2 = 0} = \frac{1}{2} [f'(-\epsilon) - f'(\epsilon)] \tag{7-51}$$

so that the efficacy becomes

$$\xi_{DZE} = \frac{[f'(-\epsilon) - f'(\epsilon)]^2}{8[1 - F(\epsilon)][2F(\epsilon) - 1]} \tag{7-52}$$

For the unit-variance *double-exponential* noise density function of (7-27) we get

$$f'(-\epsilon) - f'(\epsilon) = 2\sqrt{2}\, f(\epsilon) \qquad (7\text{-}53)$$

so that for small ϵ

$$\xi_{DZE} \approx \frac{f^2(0)}{F(\epsilon) - \frac{1}{2}}$$

$$\approx \frac{1}{\epsilon} f(0) \qquad (7\text{-}54)$$

Thus ξ_{DZE} for double-exponential noise can be made arbitrarily large, by making ϵ small enough. Of course, for finite sample size and non-zero signal strength there will be some non-zero optimum value of ϵ maximizing detector performance.

The above result shows that the use of the square-law detector can be very inefficient, from the point of view of asymptotic performance, for non-Gaussian noise density functions. The optimum nonlinearities e_{LO} maximizing the efficacy of (7-44) for some non-Gaussian density functions have been illustrated in the papers by Sheehy [1978], Poor and Thomas [1978b], and Lu and Eisenstein [1981]. As we would expect, for heavy-tailed noise density functions the optimum characteristic e_{LO} suppresses the effects of large-magnitude observations.

One interesting conclusion we may reach from our above discussion is that the use of a *square-law-limiter* nonlinearity

$$e_{SQL}(x) = \begin{cases} x^2, & |x| \leq a \\ a^2, & |x| > a \end{cases} \qquad (7\text{-}55)$$

should result in a good compromise performance for Gaussian noise as well as for heavier-tailed noise densities, provided that the parameter a is chosen appropriately. This will indeed result in a *robust* detector [Kassam and Poor, 1985].

Generalized Cross-Correlation Array Detector

Consider next the *generalized cross-correlation array* (GCA) statistic

$$T_{GCA}(\mathbf{X}) = 2 \sum_{j=1}^{L} \sum_{p=j+1}^{L} \left[\sum_{i=1}^{n} g(X_{ji})\, g(X_{pi}) \right] \qquad (7\text{-}56)$$

For $L = 2$ this is the GCC statistic. The characteristic g is assumed to satisfy the condition $E\{g(X_{ji}) \mid H_3\} = 0$, and g, f, and f_S assumed to have sufficient regularity to allow the following derivation to be valid. We get

$$E\{T_{GCA}(\mathbf{X}) \mid K_3\}/nL(L-1) = E\{g(X_{ji})\,g(X_{pi}) \mid K_3\}$$

$$= \int_{-\infty}^{\infty} \int_{-\infty}^{\infty} g(x_1)\,g(x_2) \int_{-\infty}^{\infty} f(x_1 - \theta s)\,f(x_2 - \theta s)\,f_S(s)\,ds\,dx_1\,dx_2$$

(7-57)

Proceeding as we did above, we obtain easily

$$\frac{d}{d(\theta)^2} E\{T_{GCA}(\mathbf{X}) \mid K_3\}\bigg|_{\theta^2=0} = nL(L-1)\left[\int_{-\infty}^{\infty} g(x)\,f'(x)\,dx\right]^2$$

(7-58)

Now

$$V\{T_{GCA}(\mathbf{X}) \mid H_3\} = 4n\ V\left\{\sum_{j=1}^{L}\sum_{p=j+1}^{L} g(X_{ji})\,g(X_{pi}) \mid H_3\right\}$$

$$= 4n\ \frac{L(L-1)}{2}\ V\{g(X_{ji})\,g(X_{pi}) \mid H_3\}$$

$$= 2nL(L-1)\left[\int_{-\infty}^{\infty} g^2(x)f(x)\,dx\right]^2$$

(7-59)

This finally gives the result for the efficacy ξ_{GCA} of the GCA detector,

$$\xi_{GCA} = \frac{1}{2} L(L-1)\ \frac{\left[\int_{-\infty}^{\infty} g(x)f'(x)\,dx\right]^4}{\left[\int_{-\infty}^{\infty} g^2(x)f(x)\,dx\right]^2}$$

(7-60)

It is interesting to note that

$$\xi_{GCA} = \frac{1}{2} L(L-1)\ \xi_{GCC}$$

$$= \frac{1}{2} L(L-1)\ \tilde{\xi}_{GC}^2$$

(7-61)

where $\tilde{\xi}_{GC}$ is the normalized efficacy of a GC detector for H_1 versus K_1, as developed in Section 2.4 of Chapter 2. Thus the efficacy ξ_{GCA} is closely related to the efficacy ξ_{GC} of a generalized correlator detector for a deterministic signal. In particular, we find that with $g \equiv g_{LO}$ in (7-60) [g_{LO} was defined in (2-33)] we get the maximum value for the efficacy ξ_{GCA}.

Generalization of the LO Detector

Consider now the generalization

$$\lambda_G(\mathbf{X}) = \sum_{j=1}^{L} T_{GE}(\mathbf{X}_j) + T_{GCA}(\mathbf{X}) \quad (7\text{-}62)$$

of the LO test statistic $\lambda_{LO}(\mathbf{X})$ of (7-23). Utilizing the above results and with simple further considerations for the variance of $\lambda_G(\mathbf{X})$ under H_3, we obtain the result for its efficacy:

$$\xi_G = \frac{L}{2} \frac{\left\{ \frac{1}{2} \int_{-\infty}^{\infty} e(x) f''(x)\, dx + (L-1)\left[\int_{-\infty}^{\infty} g(x) f'(x)\, dx\right]^2 \right\}^2}{\frac{1}{2}\int_{-\infty}^{\infty} e^2(x) f(x)\, dx + (L-1)\left[\int_{-\infty}^{\infty} g^2(x) f(x)\, dx\right]^2} \quad (7\text{-}63)$$

Using $e \equiv f''/f$ and $g \equiv g_{LO}$ we get from the above the efficacy of the LO statistic as

$$\xi_{LO} = \frac{L}{2}\left\{ \frac{1}{2}\int_{-\infty}^{\infty}\left[\frac{f''(x)}{f(x)}\right]^2 f(x)\, dx + (L-1) I^2(f) \right\} \quad (7\text{-}64)$$

where $I(f)$ is Fisher's information for location, defined by (2-12). The results (7-63) and (7-64) also indicate that for L reasonably large the optimum GCA detector has asymptotic performance which is quite close to that of the LO detector, when $\int_{\infty}^{\infty} [f''(x)/f(x)]^2 f(x)\, dx$ has a finite value.

In the special case when $L = 2$, ξ_{LO} of (7-64) becomes

$$\xi_{LO} = \frac{1}{2}\int_{-\infty}^{\infty}\left[\frac{f''(x)}{f(x)}\right]^2 f(x)\, dx + I^2(f) \quad (7\text{-}65)$$

For unit-variance *Gaussian* noise we have $I^2(f) = 1$ and

$$\frac{1}{2}\int_{-\infty}^{\infty}\left[\frac{f''(x)}{f(x)}\right]^2 f(x)\, dx = \frac{1}{2}\int_{-\infty}^{\infty}(x^4 - 2x^2 + 1) f(x)\, dx$$

$$= 1 \qquad (7\text{-}66)$$

showing that the square-law energy detector part and the linear cross-correlation array detector part of λ_{LO} contribute in equal amounts to the optimum asymptotic performance.

7.5 Asymptotically Optimum Quantization

As we have done in earlier sections on known low-pass and narrowband signal detection, we will now briefly study optimum quantization of the data in random signal detection problems. The importance of studying quantized-data systems needs no further emphasis. An additional factor contributing to this importance, which has not yet been mentioned, is that it is possible to design simple *nonparametric* detectors based on quantized data. This idea will not be explored in this book; discussion and references on this topic may be found in [Kassam, 1977a] and in the book by Kassam and Thomas [1980]. In this particular section the criteria of local and asymptotic optimality will be used to obtain optimum quantization schemes for use in two special cases of random signal detection. The first one corresponds to the situation when $L = 1$ in the model of (7-1), the single-receiver or single-channel case. The second case we will examine is that of the GCA test statistic for $L > 1$, which has been explained as being of special interest in detection and source location schemes.

Quantizer Energy Detector ($L = 1$)

In the generalized energy (GE) detector test statistic $T_{GE}(\mathbf{X}_1)$ of (7-32) let us now require the function e to be some M-interval staircase quantizer characteristic q defined by

$$q(x) = l_j \text{ for } t_{j-1} < x \leq t_j, \; j = 1, 2, ..., M \qquad (7\text{-}67)$$

with $-\infty \triangleq t_0 \leq t_1 \leq \cdots \leq t_M \triangleq \infty$. Using this function for e in the efficacy expression (7-44), the efficacy ξ_{QE} of the quantizer energy detector is obtained as

$$\xi_{QE} = \frac{1}{4} \frac{\left\{ \sum_{j=1}^{M} l_j [f'(t_j) - f'(t_{j-1})] \right\}^2}{\sum_{j=1}^{M} l_j^2 [F(t_j) - F(t_{j-1})]} \qquad (7\text{-}68)$$

provided that $q(X_{1i})$ is constrained (without loss of generality) to have mean value zero under the null hypothesis. This means that we require

$$\sum_{j=1}^{M} l_j \, [F(t_j) - F(t_{j-1})] = 0 \qquad (7\text{-}69)$$

Now the result (7-68) for ξ_{QE}, with the constraint of (7-69), is exactly of the form of that for ξ_{QC} given by (4-14), with the constraint (4-15). Thus we may conclude that the AO quantizer parameters must satisfy

$$l_j = \frac{f'(t_j) - f'(t_{j-1})}{F(t_j) - F(t_{j-1})}, \quad j = 1, 2, \ldots, M \qquad (7\text{-}70)$$

and

$$e_{LO}(t_j) = \frac{l_j + l_{j+1}}{2}, \quad j = 1, 2, \ldots, M-1 \qquad (7\text{-}71)$$

where e_{LO} is the LO nonlinearity of (7-26). The above two sets of equations have to be solved simultaneously for the AO parameters. The very similar general form of this result and the previous quantization results we have discussed is quite apparent. Once again it is possible to show that the AO levels of (7-70) are also the LO levels for a given set of breakpoints, and that the LO and AO quantizer parameters are also those which minimize the mean-squared error $e(e_{LO}, q; f)$ between $e_{LO}(X_{1i})$ and $q(X_{1i})$ under H_3. Finally, generalized M-level quantization can be considered, as was done in Section 4.3 of Chapter 4 for the known signal detection problem. Such a consideration for the random signal detection problem leads to results very similar to those for generalized M-level quantization for known signals.

For numerical results for the class of generalized Gaussian noise densities the reader is referred to a paper by Alexandrou and Poor [1980]. Let us now turn to an investigation of quantization for the GCC statistic, for which the results are somewhat more novel.

Multilevel and Quantizer Cross-Correlators

The generalized cross-correlation array test statistic of (7-56) can be viewed as a sum over distinct channel pairs of a generalized cross-correlation two-input statistic of the form of (7-34). Here we will therefore restrict attention to the two-input GCC statistic $T_{GCC}(\mathbf{X}_j, \mathbf{X}_p)$.

There are two ways of introducing data quantization into the test statistic of (7-34). The GCC function g can, in one approach, be constrained to be an M-interval quantizer function and the quantizer parameters may then be sought to result in an AO scheme. It turns out that the AO quantizer parameters are exactly those for AO quantization for the known-signal detection problem, for the same noise density function. This is a consequence of the result (7-60) and (7-61), from which ξ_{QCC}, the efficacy of the *quantizer cross-correlation* (QCC) test statistic using $g \equiv q$, can be obtained as $\xi_{QCC} = 1/2\, L\,(L - 1)\, \tilde{\xi}_{QC}^2$. Thus all the results pertaining to maximization of $\tilde{\xi}_{QC}$ in Section 4.2 of Chapter 4 are applicable in the present case. This also includes results on M-level generalized quantization.

The second approach to introducing quantization for $T_{GCC}(\mathbf{X}_j, \mathbf{X}_p)$ is more general, and is based on the interpretation that $T_{GCC}(\mathbf{X}_j, \mathbf{X}_p)$ is the sum of *weighted bivariate data*,

$$T_{GCC}(\mathbf{X}_j, \mathbf{X}_p) = T_{GCC}(\mathbf{X}_{jp})$$
$$= \sum_{i=1}^{n} v(X_{ji}, X_{pi}) \qquad (7\text{-}72)$$

with $\mathbf{X}_{jp}$ representing the vector of bivariate components $[(X_{j1}, X_{p1}), (X_{j2}, X_{p2}), ..., (X_{jn}, X_{pn})]$ and v some real-valued function of two inputs. We may then require v to be an M-level function q with constant values over subsets of $\mathbb{R}^2$. For this purpose let q be the function

$$q(x_1, x_2) = l_j \text{ for } (x_1, x_2) \in A_j, \quad j = 1, 2, ..., M \qquad (7\text{-}73)$$

where $\{A_j\}_{j=1}^{M}$ is a partition of $\mathbb{R}^2$. Then it is possible to consider optimization of the partition $\{A_j\}_{j=1}^{M}$ and the level assignments l_j.

Let the function v be such that $v(X_{ji}, X_{pi})$ has mean zero under H_3. This, as we have remarked before, does not amount to any essential loss of generality. Then an obvious extension of the proof of the result (7-60) for the efficacy of a GCA detector shows that for the test statistic of (7-72), the efficacy in testing H_3 versus K_3 is

$$\xi_{GCC} = \frac{\left[\int_{-\infty}^{\infty}\int_{-\infty}^{\infty} v(x_1,x_2)\, f'(x_1) f'(x_2)\, dx_1\, dx_2 \right]^2}{\int_{-\infty}^{\infty}\int_{-\infty}^{\infty} v^2(x_1,x_2) f(x_1) f(x_2)\, dx_1\, dx_2} \qquad (7\text{-}74)$$

Recall that with $v(x_1,x_2) = \text{sgn}(x_1,x_2) = \text{sgn}(x_1)\text{sgn}(x_2)$ in (7-72) the test statistic becomes the *polarity coincidence correlation* (PCC) detector. With the multilevel function q in place of v, it is reasonable to call the resulting statistic a *multilevel coincidence correlation* (MCC) detector. Carrying this nomenclature to its logical extension, the test statistic of (7-72) should be called a *generalized coincidence correlation* detection; the abbreviation for this is also GCC. In a coincidence correlation detector the value of the output corresponding to any input depends on the set of some input partition which the input value happens to fall in, or coincide with. In a generalized coincidence correlation detector there may be an uncountably infinite number of sets forming an input partition.

Using the result (7-74), we find that for an M-level MCC detector based on q for which $q(X_{ji}, X_{pi})$ has zero mean under H_3 the efficacy is

$$\xi_{MCC} = \frac{\left[\sum_{j=1}^{M} l_j \int_{A_j} f'(x_1) f'(x_2) \, dx_1 \, dx_2\right]^2}{\sum_{j=1}^{M} l_j^2 \int_{A_j} f(x_1) f(x_2) \, dx_1 \, dx_2} \qquad (7\text{-}75)$$

Once again we find a very close correspondence between this result and that of (4-32) for the efficacy ξ_{QC} of a quantizer correlator known-signal detector. Proceeding exactly as we did in Section 4.3 of Chapter 4, we find that the AO levels and the AO partition maximizing ξ_{MCC} can be obtained as

$$l_j = \frac{\int_{A_j} g_{LO}(x_1) g_{LO}(x_2) f(x_1) f(x_2) \, dx_1 \, dx_2}{\int_{A_j} f(x_1) f(x_2) \, dx_1 \, dx_2}, \qquad (7\text{-}76)$$

$$j = 1,2,...,M$$

and

$$A_j = \{(x_1,x_2) \mid \tilde{t}_{j-1} < g_{LO}(x_1) g_{LO}(x_2) \leq \tilde{t}_j \}, \qquad (7\text{-}77)$$

$$j = 1,2,...,M-1$$

provided that $g_{LO}(X_{ji}) g_{LO}(X_{pi})$ has zero probability of being equal to any one of the $\tilde{t}_j$. Here the $\tilde{t}_j$ are, as before, given by

$$\tilde{t}_j = \frac{l_j + l_{j+1}}{2}, \quad j = 1,2,...,M-1 \qquad (4\text{-}40)$$

where the l_j are the AO levels of (7-76). Thus the two sets of equations (7-76) and (7-77) have to be satisfied simultaneously by an AO set of levels and partition. One approach to solving these equations is to start with an initial guess for $\mathbf{l} = (l_1, l_2, \ldots, l_m)$, use it to obtain the $\tilde{t}_j$ and A_j, and then use the resulting partition to get a new result for $\mathbf{l}$, after which the process is continued until convergence is obtained.

When f is a symmetric function so that g_{LO} is an odd function we find that any subset A_j of an AO partition always contains complementary pairs (x_1, x_2) and $(-x_1, -x_2)$ of bivariate points. For symmetric f it is reasonable to require *a priori* that the partition $\{A_j\}_{j=1}^M$ be *symmetric*. This means we can consider $2m$-level partitioning with a partition $\{A_j, \ j = \pm 1, \pm 2, \ldots, \pm m\}$ such that (i) if (x_1, x_2) is in A_j, then so are (x_2, x_1) and $(-x_1, -x_2)$, (ii) if (x_1, x_2) is in A_j, then $(x_1, -x_2)$ is in A_{-j}. This particular case of symmetric partitioning was originally investigated by Shin and Kassam [1979].

The structure of the AO MCC detector for two-input random signal detection is shown in Figure 7.3, based on the result (7-77) for the optimum regions. Notice that one *single-input* quantizer acting on the product $g_{LO}(X_{1i})\, g_{LO}(X_{2i})$ is required. The structure of a QCC detector is shown in Figure 7.4 for comparison. The AO MCC detector has subsets A_j in its input partition which are bounded by level curves of the AO function $g_{LO}(x_1)\, g_{LO}(x_2)$. For unit-variance Gaussian noise, for example, these level curves are hyperbolas $x_1 x_2 =$ constant. This suggests that simple-to-implement sub-optimum MCC test statistics may be obtained by constraining the level curves to be of specified shapes; for example, in the first quadrant of $\mathbb{R}^2$ we can consider the level curves $\min\{x_1, x_2\} =$ constant, for which the best choice of the constants can be made. Such constrained MCC schemes were also considered by Shin and Kassam [1978, 1979].

Finally, as an indication of how well MCC and QCC detectors can perform, let us mention some numerical values. The optimum four-level MCC detector for Gaussian noise (unit variance) has an efficacy of 0.79; for eight levels this increases to 0.935, approaching the value of unity for the LCC detector as $M \to \infty$. For the QCC detector employing four-level quantizers at *each* input the best efficacy is 0.78; for eight-level quantizers this becomes 0.93.

7.6 Detection of Narrowband Random Signals

We will conclude our discussion of random signal detectors with a brief treatment of narrowband random signal detection in additive narrowband noise. Since we have already developed in

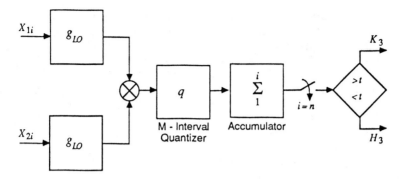

Figure 7.3 Structure of AO Two-Input Multilevel Coincidence Correlation Detector for Random Signals

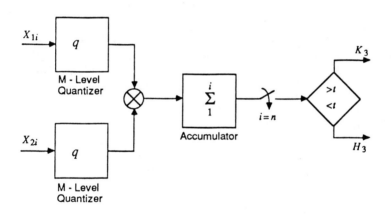

Figure 7.4 Structure of Two-Input Quantizer Cross-Correlation Detector for Random Signals

some detail the corresponding results for deterministic signal detection, we will present here only the main results and omit the details of the derivations and assumptions, which are quite similar to those we have discussed earlier.

In terms of in-phase and quadrature components, our observations are now modeled by

$$X_{Iji} = \theta S_{Ii} + W_{Iji} ,$$
$$X_{Qji} = \theta S_{Qi} + W_{Qji} ,$$
$$j = 1,2,...,L \ , \ i = 1,2,...,n \quad (7\text{-}78)$$

where the subscripts I and Q added to the quantities in (7-1) represent the in-phase and quadrature components, respectively. The bivariate density function of W_{Iji} and W_{Qji} will now be taken to be f_{IQ}. The noise in different channels, and at different sampling times in any given channel, will again be assumed to be independent. We will also make the simplifying assumption that the signal is a sequence of bivariate i.i.d. random variables. The joint density function of S_{Ii} and S_{Qi} will be denoted by f_S. It will also be assumed without giving up generality that the S_{Ii}, S_{Qi}, W_{Ii} and W_{Qi} have variances of unity, in addition to having zero means. Finally, we will assume that S_{Ii} and S_{Qi} are *uncorrelated*.

Let us focus on the case where $L = 2$, since it will be quite easy to generalize the results once the $L = 2$ case has been considered. Now the joint density function of $X_{I1i}, X_{Q1i}, X_{I2i}$, and X_{Q2i} is

$$f(x_{I1}, x_{Q1}, x_{I2}, x_{Q2} \mid \theta) = \int_{-\infty}^{\infty} \int_{-\infty}^{\infty} f_{IQ}(x_{I1} - \theta s_I, x_{Q1} - \theta s_Q)$$
$$\cdot f_{IQ}(x_{I2} - \theta s_I, x_{Q2} - \theta s_Q)$$
$$\cdot f_S(s_I, s_Q) ds_I ds_Q \qquad (7\text{-}79)$$

Then it is straightforward to show that

$$\left. \frac{\dfrac{d^2}{d\theta^2} f(x_{I1}, x_{Q1}, x_{I2}, x_{Q2} \mid \theta)}{f(x_{I1}, x_{Q1}, x_{I2}, x_{Q2} \mid \theta)} \right|_{\theta=0}$$

$$= 2 \frac{f_{IQ}^{i}(x_{I1}, x_{Q1})}{f_{IQ}(x_{I1}, x_{Q1})} \frac{f_{IQ}^{i}(x_{I2}, x_{Q2})}{f_{IQ}(x_{I2}, x_{Q2})}$$
$$+ 2 \frac{f_{IQ}^{\ii}(x_{I1}, x_{Q1})}{f_{IQ}(x_{I1}, x_{Q1})} \frac{f_{IQ}^{\ii}(x_{I2}, x_{Q2})}{f_{IQ}(x_{I2}, x_{Q2})}$$
$$+ \sum_{j=1}^{2} \left\{ \frac{f_{IQ}^{\ddot{i}}(x_{Ij}, x_{Qj})}{f_{IQ}(x_{Ij}, x_{Qj})} + \frac{f_{IQ}^{\ddot{\ii}}(x_{Ij}, x_{Qj})}{f_{IQ}(x_{Ij}, x_{Qj})} \right\} \qquad (7\text{-}80)$$

where we have used the definitions (6-11)-(6-13). Now assuming that f_{IQ} is *circularly symmetric* so that $f_{IQ}(u, v) = h(\sqrt{u^2 + v^2})$ for some function h, we obtain for the LO statistic the result

$$\lambda_{LO}(\mathbf{X}_I, \mathbf{X}_Q) = \sum_{j=1}^{2} \sum_{i=1}^{n} \left\{ \frac{h''(R_{ji})}{h(R_{ji})} + \frac{h'(R_{ji})}{R_{ji} h(R_{ji})} \right\}$$

$$+ 2 \sum_{i=1}^{n} \tilde{g}_{LO}(R_{1i}) \, \tilde{g}_{LO}(R_{2i}) \, \{X_{I1i} X_{I2i} + X_{Q1i} X_{Q2i}\}$$

(7-81)

Here the functions h and $\tilde{g}_{LO}$ are the same as those which we used in Chapters 5 and 6, and R_{ji} is the envelope of the i-th observation in the j-th array channel. Once again we note that $\lambda_{LO}(\mathbf{X}_I, \mathbf{X}_Q)$ is composed of two terms, one a sum of generalized envelope energy measures for each channel and the other a generalized narrowband cross-correlation statistic. The assumed absence of any correlation between S_{Ii} and S_{Qi} (which would be a consequence of circular symmetry for the joint density function of S_{Ii} and S_{Qi}) results in there being no cross-terms involving both X_{Iji} and X_{Qji} simultaneously in the correlation statistic. The result of (7-81) is the counterpart of the result (7-23) for low-pass random signal detection. It should be noted that the first part of the LO statistic of (7-81) is a two-input version of the LO statistic that was derived for incoherent pulse train detection in Section 6.3 of Chapter 6.

We shall not go into further details beyond this point, leaving it to the reader to develop results on efficacies and AO quantization schemes for this problem. Basically we find that the results for narrowband random signal detection can be obtained quite easily, now that we have covered the case of low-pass random signals and that of narrowband deterministic signals. Some aspects of the narrowband random signal detection problem have been treated by Levin and Rybin [1969], Korado, Zakharov, and Vinogradov [1976], and Kassam [1987].

PROBLEMS

Problem 7.1

In the observation model (7-1), replace S_i by $e_i S_i$, where the e_i are known amplitude factors not all equal to unity. Modify the derivation of the LO statistic to obtain a more general version of (7-18) for this case.

Problem 7.2

Let $L = 2$ in (7-1). For this two-input model assume that the noise terms W_{1i} and W_{2i} are uncorrelated but not independent, the assumption of temporal independence of (W_{1i}, W_{2i}), $i = 1,2,...,n$, being retained. Modify the derivation of the LO statistic to obtain a more general version of (7-18) and of (7-23) for this case.

Problem 7.3

Verify the correctness of (7-40) as an alternative version for $\lambda_{LO}(\mathbf{X})$ of (7-37).

Problem 7.4

Let $X_j(t)$, $j = 1,2$, be continuous-time observation processes with
$$X_j(t) = \theta S(t) + W_j(t)$$
in which $S(t)$, $W_1(t)$, and $W_2(t)$ are zero-mean independent random processes. Let $\Phi_S(\omega)$ be the power spectral density (psd) of $S(t)$, and let $\Phi_W(\omega)$ be the common psd of the $W_j(t)$. Consider a detector in which the process $X_1(t)$ is passed through a filter with frequency response $H(\omega)$ to produce $Y_1(t)$, the process $X_2(t)$ is passed through a filter with frequency response $H^*(\omega)$ to produce $Y_2(t)$, and the outputs $Y_1(t)$ and $Y_2(t)$ are multiplied and the product integrated over a time interval of length T. Let the result be the quantity $Z(T)$.

Consider the SNR criterion
$$D_Z = \frac{[E\{Z(T)\}]^2}{V\{Z(T) \mid \theta^2 = 0\}}$$
in the asymptotic case $T \to \infty$, where in evaluating the numerator we let $\theta^2 \to 0$ such that $\theta^2 \sqrt{T} \to \gamma$, a finite, positive constant. Show that the Eckart filter with frequency response
$$H(\omega) = \frac{\sqrt{\Phi_S(\omega)}}{\Phi_W(\omega)}$$
maximizes D_Z.

Problem 7.5

Change the model of (7-1) to the following one for the X_{ji}:
$$X_{ji} = \theta^2 \mu_i + \theta S_i + W_{ji}$$
for $j = 1,2,...,L$ and $i = 1,2,...,n$, with $\{S_i\}_{i=1}^n$ now being a zero-mean process and the μ_i being known constants. Otherwise, the three assumptions following (7-1) are retained. Find the general form of the LO statistic for testing $\theta = 0$ versus $\theta \neq 0$ in this situation.

Problem 7.6

In the model (7-1) for the X_{ji} let $L = 1$ and let the S_i be independent random variables for $i = 1,2,...,n$. Suppose now that we can express S_i as $S_i = \mu_i + N_i$, where the μ_i are known mean values and $\{N_i\}_{i=1}^n$ is a sequence of zero-mean independent and identically distributed random variables, with N_i and W_i generally *dependent* random variables. Let f_{WN} be the joint pdf of W_i and N_i. The W_i are independent and identically distributed zero-mean, unit-variance noise components as before. Obtain the LO test statistic for $\theta = 0$ versus $\theta > 0$ in this model.

Consider the special case in which $E\{N_i \mid W_i\} = \rho W_i$. Find in this case the LO statistic when (a) W_i is Gaussian, and (b) W_i has the double-exponential pdf.

Problem 7.7

Obtain the ARE of the LO detector based on the statistic derived in Problem 7.6 relative to the LO detector obtained for independent N_i and W_i, when W_i and N_i have joint pdf f_{WN}. Evaluate this ARE as a function of ρ for $E\{N_i \mid W_i\} = \rho W_i$ when (a) W_i is Gaussian, and (b) W_i has the double-exponential pdf.

Problem 7.8

Carry out a finite-sample-size performance study of the dead-zone generalized energy detector [which uses the test statistic of (7-32) with $e(x) = d_\epsilon(x)$ of (7-31)], when f is the unit-variance double-exponential noise pdf and θS_i has the pdf

$$f_{\theta S}(x) = \frac{1}{2}\delta(x-s) + \frac{1}{2}\delta(x+s)$$

where δ is the unit-impulse or delta function. You should characterize its detection performance for fixed false-alarm probability (say 10^{-3}) as a function of ϵ for fixed s (say 0.1 and 0.05) for a set of values of n. Compare this performance against that of the square-law detector; use numerical computations or simulations.

Problem 7.9

In Section 7.6, let $\tilde{X}_{ki} = X_{Iki} + j\, X_{Qki}$ so that $R_{ki} = |\tilde{X}_{ki}|$, and consider *amplitude hard limiting* so that $\tilde{X}_{ki}$ is converted to $\tilde{U}_{ki} = \tilde{X}_{ki}/|\tilde{X}_{ki}|$. The statistic of (7-81) with the $\tilde{X}_{ji}$ replaced by $\tilde{U}_{ji}$ is then equivalent to the *hard-limiter narrowband cross-correlation* statistic

$$T_{HNCC}(\tilde{\mathbf{X}}_1,\tilde{\mathbf{X}}_2) = \sum_{i=1}^n \text{Re}\{\tilde{U}_{1i}\,\tilde{U}_{2i}^*\}$$

Show that in testing $\theta = 0$ versus $\theta \neq 0$ in (7-78), the above test statistic allows any desired false-alarm probability to be maintained for any circularly symmetric pdf f_{IQ} of the (W_{Iji}, W_{Qji}). Derive its efficacy in this application, and obtain also the efficacy of the *linear narrowband cross-correlation* detector based on the statistic

$$T_{LNCC}(\tilde{\mathbf{X}}_1,\tilde{\mathbf{X}}_2) = \sum_{i=1}^{n} \text{Re}\{\tilde{X}_{1i}\tilde{X}_{2i}^*\}$$

for this problem.

REFERENCES

Aazhang, B., and H.V. Poor (1984), "On optimum and nearly optimum data quantization for signal detection," *IEEE Trans. Communications*, **32**, 745-751

Aazhang, B., and H.V. Poor (1985), "Asymptotic and approximate results on the performance of DS/SSMA receivers in non-Gaussian noise," *Proc. Conf. Information Sciences and Systems (Johns Hopkins Univ.)*, 520-525

Aazhang, B., and H.V. Poor (1986), "A comparison of linear and hard-limiting correlation receivers for DS/SSMA communications in impulsive channels," *Proc. Conf. Information Sciences and Systems (Princeton Univ.)*, 387-392

Abaya, E.F., and G.L. Wise (1981), "Some notes on optimal quantization," *Proc. IEEE Internat. Communications Conf.*, 30.7.1-30.7.5

Adams, S.L., and L.W. Nolte (1975), "Optimum array detection in fluctuating ambient noise," *J. Acoustical Society of America*, **58**, 670-678

Alexandrou, D., and H.V. Poor (1980), "The analysis and design of data quantization schemes for stochastic-signal detection systems," *IEEE Trans. Communications*, **28**, 983-991

Algazi, V.R., and R.M. Lerner (1964), "Binary detection in white non-Gaussian noise," *Technical Report LL/DS-2138*, M.I.T. Lincoln Laboratory, Lexington, MA

Ali, S.M., and S.D. Silvey (1966), "A general class of coefficients of divergence of one distribution from another," *J. Royal Statistical Society*, **28**, 131-142

Ammar, M., and Y.F. Huang (1985), "Quantization for signal detection using statistical moments," *Proc. Conf. Information Sciences and Systems (Johns Hopkins Univ.)*, 405-410

Antonov, O.Ye. (1967a), "Optimum detection of signals in non-Gaussian noise," *Radio Engineering and Electronic Physics*, **12**, No. 4, 541-548

Antonov, O.Ye. (1967b), "Optimum detection of signals in non-

Gaussian noise. Detection of a signal of unknown amplitude and phase," *Radio Engineering and Electronic Physics*, **12**, No. 5, 727-735

Baronkin, V.M. (1972), "Asymptotically optimal grouping of observations," *Radio Engineering and Electronic Physics*, **17**, No. 9, 1572-1576

Bello, P.A., and R. Esposito (1969), "A new method for calculating probabilities of errors due to impulsive noise," *IEEE Trans. Communication Tech.*, **17**, 368-379

Bello, P.A., and R. Esposito (1971), "Error probabilities due to impulsive noise in linear and hard-limited DPSK systems," *IEEE Trans. Communication Tech.*, **19**, 14-20

Benitz, G.R., and J.A. Bucklew (1986), "The asymptotic theory of quantization degradation of detector performance on IID data," *Proc. Conf. Information Sciences and Systems (Princeton Univ.)*, 6-10

Bernstein, S.L., M.L. Burrows, J.E. Evans, A.S. Griffiths, D.A. McNeill, C.W. Niessen, I. Richer, D.P. White, and D.K. William (1974), "Long-range communications at extremely low frequencies," *Proceedings IEEE*, **62**, 292-312

Bernstein, S.L., D.A. McNeill, and I. Richer (1974), "A signaling scheme and experimental receiver for extremely low frequency (ELF) communication," *IEEE Trans. Communications*, **22**, 508-528

Bickel, P.J., and K.A. Doksum (1977), *Mathematical Statistics*, Holden-Day, San Francisco

Bushnell, W.J., and L. Kurz (1974), "The optimization and performance of detectors based on partition tests," *Proc. 12th Annual Allerton Conf. Circuits and Systems*, 1016-1023

Capon, J. (1961), "On the asymptotic efficiency of locally optimum detectors," *IRE Trans. Information Theory*, **7**, 67-71

Chelyshev, V.D. (1973), "Limiting sensitivity of reception in atmospheric noise conditions," *Telecommunications and Radio Engineering*, **28**, No. 12, 82-85

Chen, C.-T., and S.A. Kassam (1983), "Optimum quantization of FIR Wiener and matched filters," *Proc. IEEE Internat. Communications Conf.*, F6.1.1-F61.4

Ching, Y.-C., and L. Kurz (1972), "Nonparametric detectors based on m-interval partitioning," *IEEE Trans. Information Theory*, **18**, 251-257

Cimini, L.J., and S.A. Kassam (1983), "Data quantization for narrowband signal detection," *IEEE Trans. Aerospace and Electronic Systems*, **19**, 848-858

Czarnecki, S.V., and J.B. Thomas (1983), "An adaptive detector for non-Gaussian noise," *Proc. IEEE Internat. Communications Conf.*, F6.6.1-F6.6.5

Czarnecki, S.V., and K.S. Vastola (1985), "Approximation of locally optimum detector nonlinearities," *IEEE Trans. Information Theory*, **31**, 835-838

Dadi, M.I., and R.J. Marks (1987), "Detector relative efficiencies in the presence of Laplace noise," *IEEE Trans. Aerospace and Electronic Systems*, **23**

Dwyer, R.F. (1980), "Robust sequential detection of weak signals in undefined noise using acoustical arrays," *J. Acoustical Society of America*, **67**, 833-841

Enge, P.K. (1985), "Discrete time detection in m-dependent narrowband noise," *Proc. Conf. Information Sciences and Systems (Johns Hopkins Univ.)*, 398-404

Enge, P.K. (1986), "Asymptotic comparison of detectors with and without quadrature observations," *Proc. Conf. Information Sciences and Systems (Princeton Univ.)*, 197-203

Evans, J.E., and A.S. Griffiths (1974), "Design of a Sanguine noise processor based upon world-wide extremely low frequency (ELF) recordings," *IEEE Trans. Communications*, **22**, 528-539

Fedele, G., L. Izzo, and L. Paura (1984), "Optimum and suboptimum space-diversity detection of weak signals in non-Gaussian noise," *IEEE Trans. Communications*, **32**, 990-997

Ferguson, T.S. (1967), *Mathematical Statistics*, Academic Press, New York

Freeman, D.F. (1981), "Asymptotic performance of nonlinear demodulators in impulsive noise," *IEEE Trans. Communications*, **29**, 696-706

Gardner, W.A. (1980), "A unifying view of second-order measures of quality for signal classification," *IEEE Trans. Communications*, **28**, 807-816

Gardner, W.A. (1982), "Structural characterization of locally optimum detectors in terms of locally optimum estimators and correlators," *IEEE Trans. Information Theory*, **28**, 924-932

Grettenberg, T.L. (1963), "Signal selection in communication and radar systems," *IEEE Trans. Information Theory*, **9**, 265-275

Groeneveld, R.A. (1972), "Asymptotically optimal group rank tests for location," *J. American Statistical Association*, **67**, 847-849

Hajek, J., and Z. Sidak (1967), *Theory of Rank Tests*, Academic Press, New York

Hall, H.M. (1966), "A new model for 'impulsive' phenomena: application to atmospheric-noise communication channels," *Technical Reports 3412-8 and 7050-7*, Stanford Electron. Lab., Stanford Univ., Stanford, CA

Halverson, D.R., and G.L. Wise (1980a), "A detection scheme for dependent noise processes," *J. Franklin Inst.*, **309**, 287-300

Halverson, D.R., and G.L. Wise (1980b), "Discrete-time detection in ϕ-mixing noise," *IEEE Trans. Information Theory*, **26**, 189-198

Halverson, D.R., and G.L. Wise (1981), "Asymptotic memoryless detection of random signals in dependent noise," *J. Franklin Inst.*, **312**, 13-29

Halverson, D.R., and G.L. Wise (1984a), "Approximately optimal memoryless detection of random signals in dependent noise," *IEEE Trans. Information Theory*, **30**, 420-424

Halverson, D.R., and G.L. Wise (1984b), "Asymptotic memoryless discrete-time detection of ϕ-mixing signals in ϕ-mixing noise," *IEEE Trans. Information Theory*, **30**, 415-417

Halverson, D.R., and G.L. Wise (1986), "On nonlinearities in asymptotic memoryless detection," *IEEE Trans. Information Theory*, **32**, 292-296

Hansen, V.G. (1969), "Weak-signal optimization of multilevel quantization and corresponding detection performance," *NTZ Commun. J.*, **22**, 120-123

Hansen, V.G. (1974), "Optimization and performance of multilevel quantization in automatic detectors," *IEEE Trans. Aerospace and Electronic Systems*, **10**, 274-280

Hatsell, C.P., and L.W. Nolte (1971), "On transient and periodic signal detection in time-varying noise power," *IEEE Trans. Aerospace and Electronic Systems*, **7**, 1100-1112

Helstrom, C.W. (1967), "Detection theory and quantum mechanics," *Information and Control*, **10**, 254-291

Helstrom, C.W. (1968), *Statistical Theory of Signal Detection*, 2nd Ed., Pergamon Press, Oxford

Huang, Y.F., and J.B. Thomas (1983), "Signal detection in nearly Gaussian skewed noise," *J. Acoustical Society of America*, **74**, 1399-1405

Hug, H. (1980), "Suboptimum detectors for discrete-time signals in the presence of non-Gaussian noise," in *Signal Processing: Theories and Applications*, M. Kunt and F. de Coulon, Eds., North-Holland, Amsterdam, 315-320

Ingram, R.F. (1984), "Performance of the locally optimum threshold receiver and several suboptimal nonlinear receivers for ELF noise," *IEEE J. Oceanic Engineering*, **9**, 202-208

Izzo, L., and L. Paura (1986), "Asymptotically optimum space diversity detection in non-Gaussian noise," *IEEE Trans. Communications*, **34**, 97-103

Kailath, T. (1967), "The divergence and Bhattacharyya distance measures in signal selection," *IEEE Trans. Communications*, **15**, 52-60

Kanefsky, M., and J.B. Thomas (1965), "On polarity detection schemes with non-Gaussian inputs," *J. Franklin Inst.*, **280**, 120-138

Kassam, S.A. (1977a), "Conditional tests in nonparametric detection," in *Nonparametric Methods in Communications*, P. Papantoni-Kazakos and D. Kazakos, Eds., Marcel Dekker, New York, 145-201

Kassam, S.A. (1977b), "Optimum quantization for signal detection," *IEEE Trans. Communication*, **25**, 479-484

Kassam, S.A. (1978), "Locally robust array detectors for random signals," *IEEE Trans. Information Theory*, **24**, 309-316

Kassam, S.A. (1981), "The performance characteristics of two extensions of the sign detector," *IEEE Trans. Communications*, **29**, 1038-1044

Kassam, S.A. (1985), "Optimum data quantization in signal detection," in *Communications and Networks: A Survey of Recent Advances*, I.F. Blake and H.V. Poor, Eds., Springer-Verlag, New York, 72-110

Kassam, S.A. (1987), "Nonparametric hard limiting and sign detection of narrowband deterministic and random signals," *IEEE J. Oceanic Engineering*, **12**, 66-74

Kassam, S.A., and T.L. Lim (1978), "Coefficient and data quantization in matched filters for detection," *IEEE Trans. Communications*, **26**, 124-127

Kassam, S.A., and H.V. Poor (1985), "Robust techniques for signal processing: a survey," *Proceedings IEEE*, **73**, 433-481

Kassam, S.A., and J.B. Thomas (1976), "Generalizations of the sign detector based on conditional tests," *IEEE Trans. Communications*, **24**, 481-487

Kassam, S.A., and J.B. Thomas, Eds. (1980), *Nonparametric Detection Theory and Applications*, Benchmark Papers in Electrical Engineering and Computer Science, **23**, Dowden, Hutchinson, and Ross, Stroudsburg, PA (Academic Press)

Kendall, M.G., and A. Stuart (1967), *The Advanced Theory of Statistics, Vol. II*, 2nd Ed., Hafner, New York

Klose, D.R., and L. Kurz (1969), "A new representation theory and detection procedures for a class of non-Gaussian channels," *IEEE Trans. Communication Tech.*, **17**, 225-234

Kobayashi, H. (1970), "Distance measures and asymptotic relative efficiency," *IEEE Trans. Information Theory*, **16**, 288-291

Kobayashi, H., and J.B. Thomas (1967), "Distance measures and related criteria," *Proc. 5th Annual Allerton Conf. Circuits and Systems*, 491-500

Korado, V.A., S.I. Zakharov, and A.B. Vinogradov (1976), "Detection of a stochastic signal in the presence of a non-Gaussian interference," *Radio Engineering and Electronic Physics*, **21**, No.

2, 144-146

Kurz, L. (1962), "A method of digital signaling in the presence of additive Gaussian and impulsive noise," *IRE International Convention Rec.*, **10**, Part 4, 161-173

Kurz, L. (1977), "Nonparametric detectors based on partition tests," in *Nonparametric Methods in Communications*, P. Papantoni-Kazakos and D. Kazakos, Eds., Marcel Dekker, New York, 73-143

Kutoyants, Yu.A. (1975), "A class of asymptotically optimal algorithms for signal discrimination," *Radio Engineering and Electronic Physics*, **20**, No. 4, 131-133

Kutoyants, Yu.A. (1976), "On the asymptotic theory of signal detection," *Radio Engineering and Electronic Physics*, **21**, No. 7, 74-81

Lee, C.C., and J.B. Thomas (1983), "Detectors for multinomial input," *IEEE Trans. Aerospace and Electronic Systems*," **19**, 288-296

Lehmann, E.L. (1959), *Testing Statistical Hypotheses*, Wiley, New York

Levin, B.R., and V.M. Baronkin (1973), "Asymptotically optimum algorithms of detection of signals from quantized observations," *Radio Engineering and Electronic Physics*, **18**, No. 5, 682-689

Levin, B.R., and A.F. Kushnir (1969), "Asymptotically optimal algorithms of detection and extraction of signals from noise," *Radio Engineering and Electronic Physics*, **14**, No. 2, 213-221

Levin, B.R., and A.F. Kushnir, and A.I. Pinskiy (1971), "Asymptotically optimal algorithms for detection and discrimination of signals immersed in correlated noise," *Radio Engineering and Electronic Physics*, **16**, No. 5, 784-793

Levin, B.R., and A.K. Rybin (1969), "Nonparametric amplitude and phase methods of signal detection I," *Engineering Cybernetics*, **7**, No. 5, 103-110

Liu, Y.-C., and G.L. Wise (1983), "Some relative efficiency results for detectors for time varying signals," *Proc. IEEE Internat. Communications Conf.*, F6.4.1-F6.4.5

Lu, N.H., and B.A. Eisenstein (1981), "Detection of weak signals in non-Gaussian noise," *IEEE Trans. Information Theory*, **27**, 755-771

Lu, N.H., and B.A. Eisenstein (1983), "Suboptimum detection of weak signals in non-Gaussian noise," *IEEE Trans. Information Theory*, **29**, 462-466

Lu, N.H., and B.A. Eisenstein (1984), "Weak signal detection in non-Gaussian noise of unknown level," *IEEE Trans. Aerospace and Electronic Systems*, **20**, 830-834

Maras, A.M., H.D. Davidson, and A.G.J. Holt (1985a), "Resolution of binary signals for threshold detection in narrowband non-Gaussian noise," *IEE Proceedings*, **132**, Pt. F, No. 3, 187-192

Maras, A.M., H.D. Davidson, and A.G.J. Holt (1985b), "Weak-signal DPSK detection in narrow-band impulsive noise," *IEEE Trans. Communications*, **33**, 1008-1011

Marks, R.J., G.L. Wise, D.G. Haldeman, and J.L. Whited (1978), "Detection in Laplace noise," *IEEE Trans. Aerospace and Electronic Systems*, **14**, 866-872

Martinez, A.B., P.F. Swaszek, and J.B. Thomas (1984), "Locally optimal detection in multivariate non-Gaussian noise," *IEEE Trans. Information Theory*, **30**, 815-822

Max, J. (1960), "Quantizing for minimum distortion," *IRE Trans. Information Theory*, **6**, 7-12

Mertz, P. (1961), "Model of impulsive noise for data transmission," *IRE Trans. Communication Systems*, **9**, 130-137

Michalsky, D.L., G.L. Wise and H.V. Poor (1982), "A relative efficiency study of some popular detectors," *J. Franklin Inst.*, **313**, 135-148

Middleton, D. (1960), *An Introduction to Statistical Communication Theory*, McGraw-Hill, New York, 1598-1609

Middleton, D. (1966), "Canonically optimum threshold detection," *IEEE Trans. Information Theory*, **12**, 230-243

Middleton, D. (1977), "Statistical-physical models of electromagnetic interference," *IEEE Trans. Electromagnetic Compatibility*, **19**, 106-127

Middleton, D. (1979a), "Procedures for determining the

parameters of the first-order canonical models of class A and class B electromagnetic interference," *IEEE Trans. Electromagnetic Compatibility*, **21**, 190-208

Middleton, D. (1979b), "Canonical non-Gaussian noise models: their implications for measurement and for prediction of receiver performance," *IEEE Trans. Electromagnetic Compatibility*, **21**, 209-220

Middleton, D. (1983), "Canonical and quasi-canonical probability models of class A interference," *IEEE Trans. Electromagnetic Compatibility*, **25**, 76-106

Middleton, D. (1984), "Threshold detection in non-Gaussian interference environments: exposition and interpretation of new results for EMC applications," *IEEE Trans. Electromagnetic Compatibility*, **26**, 19-28

Mihajlovic, R.A., and L. Kurz (1982), "Orthogonal-mixing adaptive detection in incompletely specified noise," *Proc. IEEE Internat. Communications Conf.*, 1H.4.1-1H.4.6

Miller, J.H., and J.B. Thomas (1972), "Detectors for discrete-time signals in non-Gaussian noise," *IEEE Trans. Information Theory*, **18**, 241-250

Miller, J.H., and J.B. Thomas (1975), "Numerical results on the convergence of relative efficiencies," *IEEE Trans. Aerospace and Electronic Systems*, **11**, 204-209

Miller, J.H., and J.B. Thomas (1976), "The detection of signals in impulsive noise modeled as a mixture process," *IEEE Trans. Communications*, **24**, 559-563

Milne, A.R., and J.H. Ganton (1964), "Ambient noise under arctic-sea ice," *J. Acoustical Society of America*, **36**, 855-863

Modestino, J.W. (1975), "Locally optimum receiver structures for known signals in non-Gaussian narrowband noise," *Proc. 13th Annual Allerton Conf. Circuits and Systems*, 211-219

Modestino, J.W. (1977), "Adaptive detection of signals in impulsive noise environments," *IEEE Trans. Communications*, **25**, 1022-1027

Modestino, J.W., and A.Y. Ningo (1979), "Detection of weak signals in narrowband non-Gaussian noise," *IEEE Trans. Information Theory*, **25**, 592-600

Modestino, J.W., and B. Sankur (1974), "Performance of selected linear and nonlinear receivers in ELF noise," *Proc. Internat. Communications Conf.*, 30F-1 - 30F-5

Nirenberg, L. (1975), "Low SNR digital communication over certain additive non-Gaussian channels," *IEEE Trans. Communications*, **23**, 332-341

Nirenberg, L., and D. Middleton (1975), "Correction to 'Low SNR digital communication over certain additive non-Gaussian channels'," *IEEE Trans. Communications*, **23**, 1002

Noether, G.E. (1955), "On a theorem of Pitman," *Annals of Mathematical Statistics*, **26**, 64-68

Nutall, A.H. (1983), "Detection performance characteristics for a system with quantizers, OR-ing, and accumulator," *J. Acoustical Society of America*, **73**, 1631-1642

Pinskiy, A.I. (1971), "Asymptotic optimum algorithms for the detection of a weak signal, using a dependent observation sample," *Telecommunication and Radio Engineering*, **26**, No. 4, 77-80

Pitman, E.J.G. (1948), "Notes on nonparametric statistical inference," *Mimeographed notes*, Columbia Univ., New York

Poor, H.V. (1982), "Signal detection in the presence of weakly dependent noise - Part I: Optimum detection," *IEEE Trans. Information Theory*, **28**, 735-744

Poor, H.V., and D. Alexandrou (1980), "A general relationship between two quantizer design criteria," *IEEE Trans. Information Theory*, **26**, 210-212

Poor, H.V., and Y. Rivani (1981), "Input amplitude compression in digital signal-detection systems," *IEEE Trans. Communications*, **29**, 707-710

Poor, H.V., and J.B. Thomas (1977a), "Applications of Ali-Silvey distance measures in the design of generalized quantizers for binary decision systems," *IEEE Trans. Communications*, **25**, 893-900

Poor, H.V., and J.B. Thomas (1977b), "Optimum quantization for local decisions based on independent samples," *J. Franklin Inst.*, **303**, 549-561

Poor, H.V., and J.B. Thomas (1978a), "Asymptotically robust

quantization for detection," *IEEE Trans. Information Theory*, **24**, 222-229

Poor, H.V., and J.B. Thomas (1978b), "Locally optimum detection of discrete-time stochastic signals in non-Gaussian noise," *J. Acoustical Society of America*, **63**, 75-80

Poor, H.V., and J.B. Thomas (1979), "Memoryless discrete-time detection of a constant signal in m-dependent noise," *IEEE Trans. Information Theory*, **25**, 54-61

Poor, H.V., and J.B. Thomas (1980), "Memoryless quantizer-detectors for constant signals in m-dependent noise," *IEEE Trans. Information Theory*, **26**, 423-432

Rappaport, S.S., and L. Kurz (1966), "An optimal nonlinear detector for digital data transmission through non-Gaussian channels," *IEEE Trans. Communication Tech.*, **14**, 266-274

Rice, S.O. (1944), "Mathematical analysis of random noise," *Bell System Technical Journal*, **23**, 282-382

Richter, W.J., and T.I. Smits (1971), "Signal design and error rate for an impulse noise channel," *IEEE Trans. Communication Tech.*, **19**, 446-458

Rowe, H.E. (1974), "Extremely low frequency (ELF) communication to submarines," *IEEE Trans. Communications*, **22**, 371-385

Rudnick, P. (1961), "Likelihood detection of small signals in stationary noise," *J. Applied Physics*, **32**, 140-143

Schwartz, S.C., and K.S. Vastola (1983), "Detection of stochastic signals in narrowband non-Gaussian noise," *Proc. 22nd IEEE Conf. Decision and Control*, 1106-1109

Sheehy, J.J. (1978), "Optimum detection of signals in non-Gaussian noise," *J. Acoustical Society of America*, **63**, 81-90

Shin, J.G., and S.A. Kassam (1978), "Three-level coincidence-correlator detectors," *J. Acoustical Society of America*, **62**, 1389-1395

Shin, J.G., and S.A. Kassam (1979), "Multilevel coincidence correlators for random signal detection," *IEEE Trans. Information Theory*, **25**, 47-53

Spaulding, A.D. (1985), "Locally optimum and suboptimum

detector performance in a non-Gaussian interference environment," *IEEE Trans. Communications*, **33**, 509-517

Spaulding, A.D., and D. Middleton (1977a), "Optimum reception in an impulsive interference environment - Part I: Coherent detection," *IEEE Trans. Communications*, **25**, 910-923

Spaulding, A.D., and D. Middleton (1977b), "Optimum reception in an impulsive interference environment - Part II: Incoherent reception," *IEEE Trans. Communications*, **25**, 924-934

Spooner, R.L. (1968), "On the detection of a known signal in a non-Gaussian noise process," *J. Acoustical Society of America*, **44**, 141-147

Struzinski, W.A. (1983), "Optimizing the performance of a quantizer," *J. Acoustical Society of America*, **73**, 1643-1647

Tantaratana, S., and J.B. Thomas (1977), "Quantization for sequential signal detection," *IEEE Trans. Communications*, **25**, 696-703

Trunk, G.V. (1976), "Non-Rayleigh sea clutter: properties and detection of targets," *Report 7986*, Naval Research Lab., reprinted in *Automatic Detection and Radar Data Processing*, D.C. Schleher, Ed., Artech House, Dedham, 1980

Van Trees, H.L. (1971), *Detection, Estimation and Modulation Theory, Part III*, Wiley, New York

Varshney, P.K. (1981), "Combined quantization-detection of uncertain signals," *IEEE Trans. Information Theory*, **27**, 262-265

Vastola, K.S. (1984), "Threshold detection in narrow-band non-Gaussian noise," *IEEE Trans. Communications*, **32**, 134-139

Wong, B.C.Y., and I.F. Blake (1986), "An algorithm for the design of generalized quantizer-detectors using independent noise samples," *Proc. Conf. Information Sciences and Systems (Princeton Univ.)*, 204-209

Zachepitsky, A.A., V.M. Mareskin, and Yu.I. Pakhomov (1972), "Detection of weak signals in narrowband noises," *Radio Engineering and Electronic Physics*, **17**, No. 10, 1637-1643

INDEX

AO = Asymptotically optimum
ARE. *see* Asymptotic relative efficiency
Adaptive detection 91-94,122-123
Adaptive quantization 122-123
Ali-Silvey distance measure 116-117
Amplitude fluctuation 160,162,168,177
Approximation of LO nonlinearity 92,104-105,111-112,118-121,126,149
Array of receivers 67,185
Asymmetrical noise pdf 83-84
Asymptotic optimality in signal detection
    incoherent signal 160,162
    known narrowband signal 147-148
    known signal in white noise 39-42,45-46,65-66
        definition 41
        of LO detectors 42-45
Asymptotically optimum quantization
    known narrowband signal 144-145,149
        approximation of LO nonlinearity 149
    known signal in white noise 99-115
        as approximation of LO nonlinearity 104-105,112-114
        M-interval quantization 99-111
        M-region quantization 111-115
    noncoherent pulse train 172-182
        as approximation of LO nonlinearity 184
        in Gaussian noise 175-181
        M-interval envelope quantization 172-175
        M-region envelope quantization 181-182
    random signal 204-208
        M-region quantization 207-208
        one-input 204-205
        two-input 206-208
Asymptotic relative efficiency 10-21
    and finite sample-size performance 54-63,107-108,110,176-180
    as ratio of efficacies 16,21
    definition 14
    extended regularity conditions 17-18
    incoherent signal detection 160,162,183
    known signal detection 46-54,75,80,84,88-90
    multivariate regularity conditions 19-20
    narrowband known signal detection 143
    of median relative to mean 23
    regularity conditions 15
Asymptotic relative gain 50
Atmospheric noise 72-73,76-77
Automatic gain control 92
Bandpass matched filter. *see* Narrowband matched filter
Bandpass noise. *see* Narrowband noise
Bandpass signal. *see* Narrowband signal
Bayes detection 64-67,146-148
    asymptotically optimum 65-67,147-148
    locally optimum 64-67,146-147
Bayes risk 9
Bayes rule 10
Bayes test 8-10
    locally optimum 10
Bhattacharrya distance 117
Bivariate pdf 130,132-133,153,165,210
Cauchy density function 23,71,79

Circularly symmetric pdf 69,132-133,153,165
Common signal detection 185,193
Consistent test 13,42
Correlation detection 193
Correlated signal 186-187
   LO detection 189-191,196-198
Data reduction 99
Dead-zone limiter 62,70-71
   bivariate 184
   for random signal 194,213
Dead-zone limiter energy detector 194,213
   efficacy 200-201
Decision theory 8-9
Density function. *see* Probability density function
Detection probability 3
Detector
   asymptotically optimum. *see* Asymptotic optimality in signal detection
   Bayes. *see* Bayes detection
   common signal 185
   correlation 193
   energy 193
   envelope. *see* Envelope detector
   locally optimum. *see* under locally optimum detection
   nonparametric 38,56,137,196,204
   quantized data 97-99 (*see also* under specific quantizer detectors)
   robust 38,90,95-96,201
   see specific entries for the following detectors:
      dead-zone limiter energy
      generalized coincidence correlation
      generalized correlator
      generalized cross-correlation
      generalized energy
      generalized envelope correlator
      generalized narrowband correlator
      hard-limiter cross-correlation
      hard-limiter narrowband correlator
      hard-limiter narrowband cross-correlation
      linear
      linear correlator
      linear cross-correlation
      linear envelope
      linear narrowband correlator
      linear narrowband cross-correlation
      multilevel coincidence correlation
      polarity coincidence correlation
      quadrature hard-limiter narrowband correlator
      quantizer correlator
      quantizer cross-correlation
      quantizer energy
      quantizer envelope correlator
      quantizer narrowband correlator
      sign
      sign correlator
      square-law
      square-law envelope
      square-law envelope correlator
   see also Test
Differential SNR 54
Distance measures 116-117
   Ali-Silvey 116-117
   Bhattacharrya 117
   J-divergence 117
Double-exponential density function 31
   in known signal detection 31-33,38,60,106
   in random signal detection 194-195,200-201
Eckart filter 188,197-198,212
Efficacy 16,18,47-54
   generalized 21,160
   normalized 52,105,141,169
   see also under specific detectors
Energy detection 193-194

quantizer 204-205
Envelope detector
  for incoherent signal 158-159
  for noncoherent pulse train 166-168,179
Envelope limiter 168,170,183-184
Envelope nonlinearity 136,146, 168,170
Envelope power function 35,39
Envelope quantization 144-145, 172-181
  in Gaussian noise 175-181
Extremely low frequency (ELF) noise 72-73,76-77
Fading 160,162
False-alarm probability 3
Fast fading 160
Fisher's information 30,69,171
  generalized Cauchy noise 82
  generalized Gaussian noise 75
Gaussian mixture density 85-86
Generalized Cauchy noise 78-79, 81
  Fisher's information 82
  LO nonlinearity (known signal detection) 81-83
Generalized Correlator detector 37,119-124,168
  adaptive 91-92,122-124
  approximation of LO nonlinearity 118-121
  assumptions 47-48
  asymptotically optimum coefficients 123-124
  coefficient quantization 124
  complex data 168
  efficacy 51-52,54,69,71,92,119, 169,171
  multivariate 69
  quantized. see Quantizer correlator detector
Generalized coincidence correlation detector 207
Generalized cross-correlation detector 193,195,201
  array 201
  efficacy 202
  narrowband signal 210-211

quantized 206,209
Generalized efficacy 21,160
Generalized energy detector 193-194,211
  efficacy 198-199
  quantized 204
Generalized envelope correlator detector
  for incoherent signal 158
  for noncoherent pulse-train 167
  quantized. see Quantizer envelope correlator detector
Generalized Gaussian noise 73-74, 78,94
  as limit of generalized Cauchy noise 80
  Fisher's information 75
  LO nonlinearity (known signal detection) 74,79
Generalized likelihood ratio test 5,32-33
Generalized narrowband correlator detector 134-137
  efficacy 140-143
  envelope nonlinearity implementation 136-137,146
  phase quantization 149-150
  quadrature detector 158-159
Generalized quantization. see M-region quantization
Generalized Rayleigh noise 137-138
  efficacies of GNC detectors 142-144
  LO nonlinearity (coherent signal) 138
Hall model 79
Hard limiter. see Sign correlator detector, Sign detector
Hard-limiter cross-correlation detector. see Polarity coincidence correlator
Hard-limiter narrowband correlator detector 135-137
  as nonparametric detector 137
  as LO detector 137
  efficacy 142-143
  phase quantization 149-150

quadrature detector 158-159,183
Hard-limiter narrowband cross-correlation detector 213
Hole-puncher function 71
Hyperbolic secant pdf 57
Hypothesis
  alternative 2
  composite 2
  null 2
  parametric 2,30-31
  simple 2
  two-sided
Hypothesis testing 2-10
  binary 2
  scale change 178,188
  sequence of problems 12-13,40
Impulsive noise 72-73,76,79,85-86,90-91
In-phase and quadrature components 127-130,152,209-210
  complex representation 146,159
  see also narrowband noise, observation process, signal
Incoherent signal 151
  in-phase and quadrature representation 152
Incoherent signal detection. see Narrowband signal detection
J-divergence 117
Known-signal detection in white noise
  ARE of detectors 46-54,75,80,84,88-90,107-108
  array of receivers 189
  as test of hypotheses 30-31,40
  assumptions 29-31,40
  Bayes 64-67
  finite sample-size performance 54-67,107-108,110
    relative efficiency 56-67
  likelihood ratio 31-32
  multivariate signal 67-69
  pulse train 25-29
  quantization 99-118
  see also LO detection
LO = Locally optimum
LeCam's lemmas 45
Likelihood ratio 4

weak-signal approximation 36-37,64-65,171,191-192
Likelihood ratio test 5 (see also Generalized likelihood ratio test)
Limiter
  dead-zone 62,70-71,184,194,213
  envelope 168,170,183-184
  hard. see Sign and sign correlator detector
  soft 70,95-96
Linear correlator detector 38
  ARE 53,75,80,84,88-90
  efficacy 53,75
Linear cross-correlation detector 195
Linear detector 38
  finite sample-size performance 46-47,55-56,58,110
  power function 55-56,58
  relative efficiency 56-63
Linear envelope detector 178-179
  finite sample-size performance 179
Linear narrowband correlator detector 134
  efficacy 142
Linear narrowband cross-correlation detector 214
Local optimality 5-8
Locally optimum detection in narrowband noise
  incoherent signal 153-161
    efficacy and ARE 160-162
    structure of LO detector 158-161
  known narrowband signal 131-139
    Bayes detection 146-148
    circularly symmetric noise pdf 132-135
    efficacy and ARE 140-143
  noncoherent pulse train 166-168
    efficacy and ARE 168-169,171
  random narrowband signal 208-211,213-214
Locally optimum detection in white noise

known signal 34-37, 64-69, 91-92
 array case 189
 asymptotic optimality of 42-46
 Bayes detection 64-67
 efficacy and ARE 75, 80, 84
 multivariate signal 67-69, 132
 random signal 188-198
  correlated signal 188-192, 196-198
  efficacy and ARE 203
  white signal 192-196
Locally optimum nonlinearity
 incoherent signal 158-159
 known signal in white noise 36-37, 68
  ELF noise 73, 77
  generalized Cauchy noise 81-83
  generalized Gaussian noise 74, 79
  mixture noise 87, 89
  optimum approximation of in GC detector 118-122
 narrowband known signal 132, 146-147
  circularly symmetric noise 133-134, 147
  generalized Rayleigh noise envelope 138
 narrowband random signal 210-211
 noncoherent pulse train 166-168, 170
 random signal in white noise 192-195
Locally optimum quantization
 known narrowband signal 144-146
 known signal 99-102, 111-113
 noncoherent pulse-train 172-174, 181-182
 random signal 205, 207
Logistic density 95
Low-pass signal 127
Low-pass signal detection. *see* Known-signal detection in white noise

M-interval envelope quantization
 coherent signal 144-146, 149
 noncoherent pulse train 172-175
  Gaussian noise 175-181
M-interval partitioning 111
M-interval quantization 97-99
(*see also* AO quantization, LO quantization)
M-level quantization 111-112
M-region envelope quantization 181-182
M-region partitioning 111
M-region quantization 111-112
 asymptotically optimum regions. *see* AO quantization
 locally optimum levels. *see* LO quantization
Matched filter
 incoherent signal 159
 known low-pass pulse-train 26-29
 known narrowband signal 134
 noncoherent pulse train 164
Max quantizer 97
Maximum distance quantization 115-118
Middleton class A noise 86-87
Minimum distortion quantization 99, 106-109
 numerical values 109
Minimum mean-square error quantization 97-99
Mixture noise density 84-90
 for noise envelope 149
 in minimax robustness 90, 95-96
 LO nonlinearity (known signal) 87, 89
Most powerful test 3-4
Multi-input detection 67, 144, 162, 185-188, 208-211
Multilevel coincidence correlation detector 207-209
 AO detector 207
 efficacy 207
Multilevel cross-correlation detector. *see* Multilevel coincidence correlation
Narrowband matched filter 134,

159,164
Narrowband noise
  bivariate pdf 130,132-133
  in-phase and quadrature components 127-130
Narrowband observation process 127-130,151-152,163-166,208-210
  in-phase and quadrature representation 128-129,152,164-165
  nonlinear envelope transformation 136-137,146
Narrowband signal
  incoherent 151-152
  known 127-128
  matched filter 134,159,164
  noncoherent pulse train 151, 163-164
  partially coherent 152
  random 208-210
  random amplitude 160,162,168, 177
  random phase 151-152,163
Narrowband signal signal detection
  incoherent signal 151-162
    ARE of detectors 160,162,183
    as test of hypotheses 152
    assumptions 152-153
    asymptotically optimum detector 160,162
    locally optimum detector 158-161
    matched filter 159
    random amplitude 160,162, 177
  known signal 127-150
    ARE of detectors 143
    as test of hypotheses 130
    assumptions 128-130
    Bayes detection 146-148
    envelope quantization 144-146
    locally optimum detector 131-139
    matched filter 134
    random amplitude 144
  noncoherent pulse train 162-172
    assumptions 162-166
    efficacy and ARE 168-169,171
    envelope quantization 172-182
    locally optimum detector 166-168,170
    quadrature matched filter 164
    random amplitude 168
  random signal 208-211
Neyman Pearson criterion 24
Neyman-Pearson lemma 4
  generalization of 6,35-36
Noise
  correlated 46,197-198
  narrowband 127-130
Non-Gaussian noise characteristics 72-73,76-77
Noncoherent pulse train 151,163-164
Nonparametric detection 38,56, 137,196,204
Orthonormal functions 93-94,121
Partially coherent signal 152
Phase quantization 149-150,183
Poisson process 85-86
Polarity coincidence correlator 196,207
Power function 2,34-35
  envelope 35,39
  of linear detector 55-56,58
  of sign detector 55-56,58
Prefiltering 90,164
Probability density function
  assumptions for noise 29-31
  asymmetrical 83-84
  atmospheric noise 72-73,76-77
  bivariate 130,132-133,153,165, 210
  Cauchy 23,71,79
  circularly symmetric 69,132-133,153
  double exponential 31
  Gaussian mixture 85-86
  generalized Cauchy 78-79,81
  generalized Gaussian 73-74,78, 94
  generalized Rayleigh 137-138
  Hall model 79
  hyperbolic secant 57

logistic 95
mixture noise 84,87,89,149
Rayleigh 137-138
Rician 178
student-t 80
Probability of detection 3
Pulse-train
 known low-pass 25-26
 noncoherent 151,163-164
Quadrature component. *see* In-phase and quadrature components
Quadrature generalized narrowband correlator detector 159
Quadrature hard-limiter narrowband correlator detector 159
 phase quantization 183
Quadrature matched filter 159,164
Quantizer characteristic 98,108,112
Quantizer correlator detector 62,102
 asymptotically optimum 102-104
 efficacy 103,105,107-109
 finite sample size performance 110
 locally optimum 101-102
Quantizer cross-correlation detector 205-206,209
Quantized data 97-99,111-112, 144-145,172,204
 bivariate 181-182,184,206
 phase quantization 149-150,183
Quantizer energy detector 204-205
 efficacy 204
 locally optimum 205
Quantizer envelope correlator detector 173-174
 asymptotic loss 176-177
 asymptotically optimum 175
 efficacy 173-174
 finite sample-size performance 179-180
 Gaussian noise 175-180
 locally optimum 173
Quantizer narrowband correlator detector 144

efficacy 145
 phase quantization 149-150,183
Radar clutter 85
Random phase signal 151-152,163
Random power level Gaussian noise 83,94-95
Random signal detection
 as test of hypotheses 187-188
 assumptions 185-187
 correlated noise 197-198
 correlated signal 196-197
 dead-zone limiter 194
 efficacy 198-204
 locally optimum 188-198
 narrowband signal 208-211
 quantization 204-208
 scale-change model 188
 white signal 192-196
Randomization 4
Randomized decision rule 9
Rayleigh amplitude fluctuation 177
Rayleigh noise pdf 137-138
Receiver operating characteristic 46
Relative efficiency 11-12
 linear and sign detectors 56-63
Rician pdf 178
Risk function 9
Robust detection 38,90,95-96,201
Sample median 23
Sampled signal 24-25,127-130
Sign correlator detector 38
 ARE 53,75,80,88-90
 efficacy 53
Sign detector 38
 finite sample size performance 55-56,110
 nonparametric 38,56
 power function 55-56,58
 relative efficiency 56-63
Signal
 assumptions 40,128,152,163-164, 186-188
 energy 40
 incoherent 151-152
 lowpass 127
 multivariate 67

narrowband. *see* Narrowband signal
power 40,140-141
pulse-train 25-26,163-164
sampled 24-25,127-130
Signal to noise ratio 26
  detector output 53
  differential 54
  two-input detection 212
Size 3
  asymptotic 13
Skewed distribution 83-84
Slow fading 160
Soft-limiter 70,95-96
Sonar 185
Square-law detector 194
  as UMP detector 194
  efficacy 199-200
Square-law envelope correlator statistic 167
Square-law envelope detector
  incoherent signal 158-159
  noncoherent pulse train 178
    as UMP detector 178
    finite sample-size performance 180
Square-law limiter 201
Student-t density 80
Test
  Bayes 8-10
  consistent 13
  generalized likelihood ratio 5, 32-33
  invariant 5
  locally most powerful 5
  locally optimum 5-8
  most powerful 3-4
  randomized 2
  unbiased 5,7-8
  uniformly most powerful 4
Test function 2,10-11,32
Test statistic 10
Threshold 4,11,32
Two-input detector 205-206,212
  bivariate weighting 206
  quantization 206-207
Type I error 3
Unbiasedness 7-8

Uniform quantization 115,125
Weak-signal detection
  likelihood ratio approximation 36-37,64-65,171,191-192
  see also locally optimum detection
Whitening filter 197-198